Universitext: Tracts in Mathematics

Springer
*New York
Berlin
Heidelberg
Barcelona
Budapest
Hong Kong
London
Milan
Paris
Santa Clara
Singapore
Tokyo*

Universitext

Editors (North America): S. Axler, F.W. Gehring, and P.R. Halmos

Aksoy/Khamsi: Nonstandard Methods in Fixed Point Theory
Andersson: Topics in Complex Analysis
Aupetit: A Primer on Spectral Theory
Booss/Bleecker: Topology and Analysis
Borkar: Probability Theory: An Advanced Course
Carleson/Gamelin: Complex Dynamics
Cecil: Lie Sphere Geometry: With Applications to Submanifolds
Chae: Lebesgue Integration (2nd ed.)
Charlap: Bieberbach Groups and Flat Manifolds
Chern: Complex Manifolds Without Potential Theory
Cohn: A Classical Invitation to Algebraic Numbers and Class Fields
Curtis: Abstract Linear Algebra
Curtis: Matrix Groups
DiBenedetto: Degenerate Parabolic Equations
Dimca: Singularities and Topology of Hypersurfaces
Edwards: A Formal Background to Mathematics I a/b
Edwards: A Formal Background to Mathematics II a/b
Foulds: Graph Theory Applications
Fuhrmann: A Polynomial Approach to Linear Algebra
Gardiner: A First Course in Group Theory
Gårding/Tambour: Algebra for Computer Science
Goldblatt: Orthogonality and Spacetime Geometry
Gustafson/Rao: Numerical Range: The Field of Values of Linear Operators and
Matrices
Hahn: Quadratic Algebras, Clifford Algebras, and Arithmetic Witt Groups
Holmgren: A First Course in Discrete Dynamical Systems
Howe/Tan: Non-Abelian Harmonic Analysis: Applications of $SL(2, R)$
Howes: Modern Analysis and Topology
Humi/Miller: Second Course in Ordinary Differential Equations
Hurwitz/Kritikos: Lectures on Number Theory
Jennings: Modern Geometry with Applications
Jones/Morris/Pearson: Abstract Algebra and Famous Impossibilities
Kannan/Krueger: Advanced Analysis
Kelly/Matthews: The Non-Euclidean Hyperbolic Plane
Kostrikin: Introduction to Algebra
Luecking/Rubel: Complex Analysis: A Functional Analysis Approach
MacLane/Moerdijk: Sheaves in Geometry and Logic
Marcus: Number Fields
McCarthy: Introduction to Arithmetical Functions
Meyer: Essential Mathematics for Applied Fields
Mines/Richman/Ruitenburg: A Course in Constructive Algebra
Moise: Introductory Problems Course in Analysis and Topology
Morris: Introduction to Game Theory
Porter/Woods: Extensions and Absolutes of Hausdorff Spaces
Ramsay/Richtmyer: Introduction to Hyperbolic Geometry
Reisel: Elementary Theory of Metric Spaces
Rickart: Natural Function Algebras

(continued after index)

Mats Andersson

Topics in
Complex Analysis

 Springer

Mats Andersson
Department of Mathematics
Chalmers University of Technology
Göteborg University
S-412 96 Göteborg
Sweden

Editorial Board (North America):

Mathematics Subject Classification (1991): 30-01, 30C20, 30C15, 32A35

With 4 figures.

Library of Congress Cataloging-in-Publication Data
Andersson, Mats.
    Topics in complex analysis/Mats Andersson.
        p.   cm. – (Universitext)
    Includes bibliographical references and index.
    ISBN 0-387-94754-X (soft: alk. paper)
    1. Functions of complex variables. 2. Mathematical analysis.
    I. Title.
    QA331.7.A52   1996
    515'.9–dc20   96-11793

Printed on acid-free paper.

Production managed by Francine McNeill; manufacturing supervised by Jeffrey Taub.
Photocomposed copy prepared from the author's AMS-TeX files.
Printed and bound by Braun-Brumfield, Inc., Ann Arbor, MI.
Printed in the United States of America.

9 8 7 6 5 4 3 2 1

ISBN 0-387-94754-X Springer-Verlag New York Berlin Heidelberg   SPIN 10524446

Preface

This book is an outgrowth of lectures given on several occasions at Chalmers University of Technology and Göteborg University during the last ten years. As opposed to most introductory books on complex analysis, this one assumes that the reader has previous knowledge of basic real analysis. This makes it possible to follow a rather quick route through the most fundamental material on the subject in order to move ahead to reach some classical highlights (such as Fatou theorems and some Nevanlinna theory), as well as some more recent topics (for example, the corona theorem and the H^1-BMO duality) within the time frame of a one-semester course. Sections 3 and 4 in Chapter 2, Sections 5 and 6 in Chapter 3, Section 3 in Chapter 5, and Section 4 in Chapter 7 were not contained in my original lecture notes and therefore might be considered special topics. In addition, they are completely independent and can be omitted with no loss of continuity.

The order of the topics in the exposition coincides to a large degree with historical developments. The first five chapters essentially deal with theory developed in the nineteenth century, whereas the remaining chapters contain material from the early twentieth century up to the 1980s.

Choosing methods of presentation and proofs is a delicate task. My aim has been to point out connections with real analysis and harmonic analysis, while at the same time treating classical complex function theory. I also have tried to present some general tools that can be of use in other areas of analysis. Whereas these various aims sometimes can be incompatible, at times the scope of the book imposes some natural restrictions. For example, Runge's theorem is proved by the "Hahn–Banach method," partly because it is probably the simplest way to do so, but also because it is a technique that every student in analysis should become familiar with. However, a constructive proof is outlined as an exercise. Complex analysis is one of the origins of harmonic analysis, and several results in the latter subject have forerunners in complex analysis. Fatou's theorem in Chapter 6 is proved using standard harmonic analysis, in particular using the weak-type estimate for the Hardy–Littlewood maximal function. However, most standard tools from harmonic analysis are beyond the scope of this book, and therefore, the L^p-boundedness of the Hilbert transform and the H^p-space theory, for instance, are treated with complex analytic methods. Carleson's inequality is proved by an elementary argument due to B. Berndtsson, rather than using the L^p-estimate for the maximal function, and Carleson's interpolation theorem is proved using the beautiful and explicit construction of the interpolating function due to P. Jones from the 1980s. However, a proof based on the $\bar{\partial}_b$-equation is indicated in an exercise.

Necessary prerequisites for the reader are basic courses in integration theory and functional analysis. In the text, I sometimes refer to distribution theory, but this is merely for illustration and can be skipped over with no serious loss of understanding. The reader whose memory of an elementary (undergraduate) course in complex analysis is not so strong is advised to consult an appropriate text for supplementary reading.

As usual, the exercises can be divided into two categories: those that merely test the reader's understanding of or shed light on definitions and theorems (these are sometimes interposed in the text) and those that ask the reader to apply the theory or to develop it further. I think that for optimal results a good deal of the time reserved for the study of this subject should be devoted to grappling with the exercises. The exercises follow the approximate order of topics in the corresponding chapters, and thus, the degree of difficulty can vary greatly. For some of the exercises, I have supplied hints and answers.

At the end of each chapter, I have included references to the main results, usually to some more encyclopedic treatment of the subject in question, but sometimes to original papers. If references do not always appear, this is solely for the sake of expediency and does not imply any claim of originality on my part. My contribution consists mainly in the disposition and adaptation of some material and proofs, previously found only in papers or encyclopedic texts addressed to experts, into a form that hopefully will be accessible to students.

Finally, I would like to take this opportunity to express my appreciation to all of the students and colleagues who have pointed out errors and obscurities in various earlier versions of the manuscript and made valuable suggestions for improvements. For their help with the final version, I would like to thank in particular Lars Alexandersson, Bo Berndtsson, Hasse Carlsson, Niklas Lindholm, and Jeffrey Steif.

Göteborg, Sweden   Mats Andersson

Contents

Preliminaries

§1. Notation

Throughout this book the letters Ω and K always will denote open and compact sets, respectively, in $\mathbb{R}^2$, and ω will denote a bounded open set with (when necessary) piecewise C^1 boundary $\partial\omega$, which is always supposed to be positively oriented; i.e., one has ω on the left-hand side when passing along $\partial\omega$. The notation $\omega \subset\subset \Omega$ means that the closure of ω is a compact subset of Ω, and $d(K, E)$ denotes the distance between the sets K and E. Moreover, $D(a, r)$ is the open disk with center at a and radius r, and U denotes the unit disk, i.e., $U = D(0, 1)$, and T is its boundary $\partial U = \{z;\ |z| = 1\}$. The closure of a set $E \in \mathbb{R}^2$ is denoted by $\overline{E}$ and its interior is denoted by int E.

The space of k times (real) differentiable (complex valued) functions in Ω is denoted by $C^k(\Omega)$ (however, we write $C(\Omega)$ rather than $C^0(\Omega)$) and $C^\infty(\Omega) = \cap C^k(\Omega)$. Moreover, $C^k(\overline{\Omega})$ is the subspace of functions in $C^k(\Omega)$ whose derivatives up to the kth order have continuous extensions to $\overline{\Omega}$, and $C_0^k(\Omega)$ is the subspace of functions in $C^k(\Omega)$ that have compact support in Ω. Lebesgue measure in $\mathbb{R}^2$ is denoted by $d\lambda$, whereas $d\sigma$ denotes arc length along curves. We use the standard abbreviation a.e. for "almost every(where)." We also use u.c. for "uniformly on compact sets." If f, ϕ are functions, then "$f = \mathcal{O}(\phi)$ when $x \to a$" means that f/ϕ is bounded in a neighborhood of a and "$f = o(\phi)$ when $x \to a$" means that $f/\phi \to 0$ when $x \to a$. Sometimes we also use the notation $f \lesssim g$, which means that f is less than or equal to some constant times g. Moreover, $f \sim g$ stands for $f \lesssim g$ and $f \gtrsim g$.

We will use standard facts from basic courses in integration theory and functional analysis. Sometimes we also refer to distribution theory (mainly in remarks), but these comments are meant merely for illustration and always can be passed over with no loss of continuity. In the next section we have assembled some facts that will be used frequently in the text. In the first chapters we refer to them explicitly but later on often only implicitly.

Almost all necessary background material can be found in [F] or [Ru1], combined with a basic calculus book. For the facts in item B below see, e.g., [Hö], which also serves as a general reference on distribution theory.

§2. Some Facts

A. Some facts from calculus. If f is a map from Ω into $\mathbb{R}^2$ that is C^1 in a neighborhood of a, then

$$f(a + x) = f(a) + Df|_a x + o(|x|) \quad \text{when} \quad x \to 0,$$

for some linear map $x \mapsto Df|_a x$, i.e., f is *differentiable* at a. If f is considered as a complex valued function, then

$$Df|_a x = x_1 \frac{\partial f}{\partial x_1}\bigg|_a + x_2 \frac{\partial f}{\partial x_2}\bigg|_a.$$

Let $\gamma(t) = (\gamma_1(t), \gamma_2(t))$, $a \le t \le b$ be a piecewise C^1 parametrization of the curve Γ. If P, Q are continuous functions on Γ, then

$$\int_\Gamma P dx + Q dy = \int_a^b (P(\gamma_1(t), \gamma_2(t))\gamma_1'(t) + Q(\gamma_1(t), \gamma_2(t))\gamma_2'(t)) \, dt,$$

and this expression is independent of the choice of parametrization. Note that

$$\int_\Gamma f dg = \int_a^b f \circ \gamma \frac{d(g \circ \gamma)}{dt} dt$$

if $f, g \in C^1(\Omega)$ and $\Gamma \subset \Omega$. In particular, for an exact form we have

$$\int_\Gamma dg = g(\gamma(b)) - g(\gamma(a)).$$

The *arc length* of the curve Γ is

$$|\Gamma| = \int_\Gamma d\sigma = \int_a^b |\gamma'(t)|^2 dt = \int_a^b \sqrt{(\gamma_1'(t))^2 + (\gamma_2'(t))^2} dt$$

and

$$\left| \int_\Gamma P dx + Q dy \right| \le \int_\Gamma \sqrt{|P|^2 + |Q|^2} d\sigma \le |\Gamma| \sup_\Gamma \sqrt{|P|^2 + |Q|^2}.$$

Green's formula (Stokes' theorem) states that if $P, Q \in C^1(\overline{\omega})$, then

$$\int_{\partial \omega} P dx + Q dy = \int_\omega (Q_x - P_y) d\lambda,$$

whereas Green's identity (Green's formula) states that if $u, v \in C^1(\overline{\omega})$, then

$$\int_\omega (u \Delta v - v \Delta u) d\lambda = \int_{\partial \omega} \left(u \frac{\partial v}{\partial \eta} - v \frac{\partial u}{\partial \eta} \right) d\sigma,$$

where $\partial/\partial \eta$ is the outward normal derivative, i.e., $\partial u/\partial \eta = \sum \eta_j (\partial u/\partial x_j)$ if $\eta = (\eta_1, \eta_2)$ is the outward normal to $\partial \omega$.

On some occasions we also refer to the inverse function theorem, see, e.g., [Hö]: If $f: \Omega \to \mathbb{R}^2$ is C^1 and its derivative $Df|_a$ at $a \in \Omega$ is nonsingular, then locally f has a C^1 inverse g.

B. Existence of test functions. There are "plenty" of functions in $C_0^\infty(\Omega)$, namely,

(i) for any $K \subset \Omega$ there is a $\phi \in C_0^\infty(\Omega)$ such that $\phi = 1$ in a neighborhood of K and $0 \le \phi \le 1$.

(ii) if $f \in L^1_{\text{loc}}(\Omega)$ and $\int_\Omega \phi f d\lambda = 0$ for all $\phi \in C_0^\infty(\Omega)$, then $f = 0$ a.e.

(iii) if f is continuous on $K \subset \Omega$ and $\epsilon > 0$, there is a $\phi \in C_0^\infty(\Omega)$ such that $\sup_K |\phi - f| < \epsilon$.

(iv) if $\cup \Omega_\alpha = \Omega$, then there is a *smooth partition of unity subordinate to the open cover* Ω_α, i.e., there are $\phi_k \in C_0^\infty(\Omega_{\alpha_k})$ such that $0 \le \phi_k \le 1$, locally only a finite number of ϕ_k are nonvanishing and $\sum_k \phi_k \equiv 1$ in Ω.

C. Integration theory. From integration theory we use the standard convergence theorems, such as Fatou's lemma, Lebesgue's theorem on dominated convergence, and the monotone convergence theorem. Moreover, we frequently use Jensen's and Hölder's inequalities and Fubini's theorem, the duality of L^p and L^q for $p < \infty$, and the one-to-one correspondence between the continuous linear functionals on $C(K)$ and the space of finite complex Borel measures on K (usually just referred to as measures on K). Furthermore, we need the Jordan decomposition of a real measure, the Lebesgue–Radon–Nikodym decomposition of a complex measure with respect to a positive measure (the Lebesgue measure in our case) and the weak type 1-1 estimate for the Hardy–Littlewood maximal function.

In particular, we frequently will make use of "differentiation under the integral sign": Suppose that X, μ is a measure space and $f(x,t)$ is a measurable function on $X \times I$, where I is an interval in $\mathbb{R}$, which is continuously differentiable in t. Suppose further that $f(x,t)$ and $f'(x,t)$ are in $L^1(\mu)$ for each fixed t so that

$$g(t) = \int f(x,t) d\mu(x) \quad \text{and} \quad h(t) = \int f'(x,t) d\mu(x)$$

are well-defined. One may ask whether $g'(t) = h(t)$. Suppose that there is a $\psi \in L^1(\mu)$ such that

$$|f'(x,t)| \le \psi(x).$$

Then $h(t)$ is continuous by the dominated convergence theorem. Moreover,

$$\iint_{X \times I} |f'(x,t)| d\mu(x) dt < \infty,$$

so we can use Fubini's theorem:

$$\int_a^b h(t) dt = \int_a^b \left(\int f'(x,t) d\mu(x) \right) dt = \int \left(\int_a^b f'(x,t) dt \right) d\mu(x)$$

and hence

$$\int_a^b h(t)dt = \int (f(x,b) - f(x,a)) \, d\mu(x) = g(b) - g(a)$$

for all $a, b \in I$. Since h is continuous, $g' = h$.

D. Functional analysis. We will use basic results such as orthogonal decomposition and Parseval's equality in Hilbert spaces, the Hahn–Banach theorem, the Banach–Steinhaus theorem, and the open mapping theorem in Banach spaces. Moreover, on some occasions we require Arzela–Ascoli's theorem on locally equicontinuous subsets of $C(\Omega)$ and Tietze's extension theorem.

We also refer to the Fourier transform, Plancherel's formula, and the inversion formula; see, e.g., Ch. 9 in [Rul].

1

Some Basic Properties of Analytic Functions

§1. Definition and Integral Representation

We identify $\mathbb{C}$ with $\mathbb{R}^2$ by identifying the complex number $z = x + iy$ with the point $(x, y) \in \mathbb{R}^2$. Observe that a (complex-valued) differential form $Pdx + Qdy$ always can be written in the form $fdz + gd\bar{z}$, where $dz = dx + idy$ and $d\bar{z} = dx - idy$ (take $f = (P - iQ)/2$ and $g = (P + iQ)/2$). This motivates us to introduce the differential operators

$$\frac{\partial}{\partial z} = \frac{1}{2}\left(\frac{\partial}{\partial x} - i\frac{\partial}{\partial y}\right) \qquad \text{and} \qquad \frac{\partial}{\partial \bar{z}} = \frac{1}{2}\left(\frac{\partial}{\partial x} + i\frac{\partial}{\partial y}\right)$$

so that

$$df = \frac{\partial f}{\partial x}dx + \frac{\partial f}{\partial y}dy = \frac{\partial f}{\partial z}dz + \frac{\partial f}{\partial \bar{z}}d\bar{z}. \tag{1.1}$$

Note that $\Delta = \partial^2/\partial x^2 + \partial^2/\partial y^2 = 4\partial^2/\partial z\partial\bar{z}$.

1.1 Proposition. *If f is differentiable at the point a, then the limit*

$$\lim_{z \to 0} \frac{f(a + z) - f(a)}{z} \tag{1.2}$$

exists if and only if $(\partial f/\partial \bar{z})|_a = 0$; and in that case, the limit equals $(\partial f/\partial z)|_a$.

The limit (if it exists) is denoted by $f'(a)$.

Proof. The differentiability of f is equivalent to

$$f(a + z) - f(a) = z\frac{\partial f}{\partial z}\bigg|_a + \bar{z}\frac{\partial f}{\partial \bar{z}}\bigg|_a + o(|z|).$$

Thus, (1.2) exists if and only if

$$\lim_{z \to 0} \frac{\bar{z}}{z} \frac{\partial f}{\partial \bar{z}}\Big|_a$$

exists, and this holds if and only if $(\partial f / \partial \bar{z})|_a = 0$. The last statement follows immediately. $\square$

Notice that if $f = u + iv$, where u and v are real, then the *Cauchy–Riemann equation* $\partial f / \partial \bar{z} = 0$ is equivalent to

$$\begin{cases} u_x = v_y \\ u_y = -v_x \end{cases} \tag{1.3}$$

(identify the real and imaginary parts in the equation $(\partial/\partial x + i\partial/\partial y)(u + iv) = 0$).

Definition. A function $f \in C^1(\Omega)$ is *analytic* (or *holomorphic*) in Ω if $\partial f / \partial \bar{z} = 0$ in Ω. The set of analytic functions is denoted by $A(\Omega)$.

In view of Proposition 1.1, $f \in C^1(\Omega)$ is analytic in Ω if and only if (1.2) exists for all $a \in \Omega$, but we even have

1.2 Goursat's Theorem. *If f is any (complex valued) function in Ω such that (1.2) exists for all $a \in \Omega$, then f is C^1 and hence analytic.*

The proof appears later on! In most cases, it is advantageous to write the Cauchy–Riemann equation in the complex form $\partial f / \partial \bar{z} = 0$, rather than as (1.3). For instance, clearly the product rules

$$\frac{\partial}{\partial z}(fg) = \frac{\partial f}{\partial z} g + f \frac{\partial g}{\partial z} \quad \text{and} \quad \frac{\partial}{\partial \bar{z}}(fg) = \frac{\partial f}{\partial \bar{z}} g + f \frac{\partial g}{\partial \bar{z}}$$

hold; thus, if $f, g \in A(\Omega)$, one immediately finds that $fg \in A(\Omega)$ and $(fg)' = f'g + fg'$. Suppose that $h(t)$ is C^1 on an interval $I \subset \mathbb{R}$ and that f is C^1 in a neighborhood of the image of h in $\mathbb{C}$. Then by (1.1),

$$\frac{d(f \circ h)}{dt} dt = d(f \circ h) = \frac{\partial f}{\partial z} dh + \frac{\partial f}{\partial \bar{z}} d\bar{h} = \frac{\partial f}{\partial z} \frac{dh}{dt} dt + \frac{\partial f}{\partial \bar{z}} \frac{d\bar{h}}{dt} dt,$$

and therefore we have the chain rule

$$\frac{df \circ h}{dt} = \frac{\partial f}{\partial z} \frac{dh}{dt} + \frac{\partial f}{\partial \bar{z}} \frac{d\bar{h}}{dt}.$$

In the same way, if $h(\zeta)$ is C^1 in some domain in $\mathbb{C}$, then

$$\frac{\partial f \circ h}{\partial \tau} = \frac{\partial f}{\partial z} \frac{\partial h}{\partial \tau} + \frac{\partial f}{\partial \bar{z}} \frac{\partial \bar{h}}{\partial \tau} \quad \text{and} \quad \frac{\partial f \circ h}{\partial \bar{\tau}} = \frac{\partial f}{\partial z} \frac{\partial h}{\partial \bar{\tau}} + \frac{\partial f}{\partial \bar{z}} \frac{\partial \bar{h}}{\partial \bar{\tau}}.$$

Thus, if f, g are analytic, then $f \circ g$ is analytic and $(f \circ g)' = f'(g)g'$.

1.3 Example. If $f \in A(\Omega)$ and $f'(z) = 0$ for all $z \in \Omega$, then $df = 0$ and hence f is locally constant. Suppose now that $f \in A(\{\rho < |z| < R\})$ and that $f(z) = f(re^{i\theta})$ only depends on θ. Then by the chain rule

$$0 = \frac{\partial f}{\partial r} = \frac{\partial f}{\partial z}\frac{\partial z}{\partial r} + \frac{\partial f}{\partial \bar z}\frac{\partial \bar z}{\partial r} = e^{i\theta}\frac{\partial f}{\partial z} = e^{i\theta} f',$$

and therefore f is constant. The same conclusion holds if f is independent of θ.

Here are some other simple consequences of the product rule and the chain rule (and the definition).

(a) If $f, g \in A(\Omega)$ and $\alpha, \beta \in \mathbb{C}$, then $fg \in A(\Omega)$ and $\alpha f + \beta g \in A(\Omega)$.

(b) $z \mapsto z^m$, m being a natural number, is analytic in $\mathbb{C}$.

(c) If $f \in A(\Omega)$, then $1/f \in A(\Omega \setminus \{f = 0\})$. (First show that $1/z \in A(\mathbb{C} \setminus \{0\})$!)

(d) $z \mapsto e^z =_{\text{def}} e^x(\cos y + i\sin y)$ is analytic in $\mathbb{C}$.

(e) $(\partial/\partial z)z^m = mz^{m-1}$, $(\partial/\partial z)e^z = e^z$, $(\partial/\partial z)(1/f) = -f'/f^2$.

Exercise 1. Show that

(a) $\overline{\partial f/\partial \bar z} = \partial \bar f/\partial z$.

(b) if $f \in A(\Omega)$, then $z \mapsto \overline{f(\bar z)}$ is analytic in $\{z; \bar z \in \Omega\}$.

(c) if $f \in A(\Omega)$ and f is real, then f is (locally) constant.

(d) if $f \in A(\Omega)$ and $|f|$ is constant, then f is (locally) constant.

If the curve Γ is given by $r(t) = r_1(t) + ir_2(t)$, $a \le t \le b$, then, see A in the preliminaries,

$$\int_\Gamma f dz + g d\bar z = \int_a^b \left(f(r(t))r'(t) + g(r(t))\overline{r'(t)} \right) dt,$$

where of course $r'(t) = r_1'(t) + ir_2'(t)$. Moreover,

$$\left| \int_\Gamma f dz \right| = \left| \int_a^b f(r(t))r'(t)dt \right| \le \int_a^b |f(r(t))||r'(t)|dt = \int_\Gamma |f| d\sigma$$

so that $(|dz| = d\sigma)$

$$\left| \int_\Gamma f dz \right| \le \int_\Gamma |f||dz| \le |\Gamma| \sup_\Gamma |f|. \tag{1.4}$$

1.4 Proposition. If $F \in A(\Omega)$ and $f = F'$, then

$$\int_\Gamma f dz = F(r(b)) - F(r(a)).$$

In particular, $\int_\Gamma f dz = 0$ if Γ is closed.

Proof. $f dz = (\partial F/\partial z)dz = (\partial F/\partial z)dz + (\partial F/\partial \bar z)d\bar z = dF$; therefore, $f dz$ is an exact form and thus the proposition follows, cf. item A in the preliminaries. □

In complex notation Green's formula becomes (check!)

$$\int_{\partial\omega} f\,dz + g\,d\bar{z} = 2i \int_{\omega} \left(\frac{\partial f}{\partial\bar{z}} - \frac{\partial g}{\partial z}\right) d\lambda(z), \qquad f,g \in C^1(\bar{\omega}). \qquad (1.5)$$

From this we immediately get

1.5 Cauchy's Integral Theorem. *If $f \in A(\omega) \cap C^1(\bar{\omega})$, then*

$$\int_{\partial\omega} f\,dz = 0.$$

The next theorem in particular tells us that the values of an analytic function in the interior of a domain are determined by its values on the boundary.

1.6 Cauchy's Formula. *If $f \in C^1(\bar{\omega})$ and $z \in \omega$, then*

$$f(z) = \frac{1}{2\pi i} \int_{\partial\omega} \frac{f(\zeta)d\zeta}{\zeta - z} - \frac{1}{\pi} \int_{\omega} \frac{\partial f}{\partial\bar{\zeta}}(\zeta) \frac{d\lambda(\zeta)}{\zeta - z}.$$

In particular, if $f \in A(\omega) \cap C^1(\bar{\omega})$, then

$$f(z) = \frac{1}{2\pi i} \int_{\partial\omega} \frac{f(\zeta)d\zeta}{\zeta - z}.$$

Proof. Take $z \in \omega$. For ϵ so small that $\{\zeta; |\zeta - z| < \epsilon\} \subset \omega$, (1.5) (with f replaced by $f(\zeta)/(\zeta - z)$) gives that

$$2i \int_{\omega \setminus \{|\zeta - z| < \epsilon\}} \frac{\partial f}{\partial\bar{\zeta}} \frac{d\lambda(\zeta)}{\zeta - z} = \int_{\partial\omega} \frac{f\,d\zeta}{\zeta - z} - \int_{|\zeta - z| = \epsilon} \frac{f\,d\zeta}{\zeta - z}, \qquad (1.6)$$

since $\zeta \to 1/(\zeta - z)$ is analytic in $\omega \setminus \{|\zeta - z| < \epsilon|\}$. By (1.5) again, we get

$$\int_{|\zeta - z| = \epsilon} \frac{f\,d\zeta}{\zeta - z} = \frac{1}{\epsilon^2} \int_{|\zeta - z| = \epsilon} (\bar{\zeta} - \bar{z}) f\,d\zeta$$

$$= \frac{2i}{\epsilon^2} \int_{|\zeta - z| < \epsilon} \left(f(\zeta) + (\bar{\zeta} - \bar{z})\frac{\partial f}{\partial\bar{\zeta}}\right) d\lambda(\zeta)$$

$$= \frac{2\pi i}{\pi \epsilon^2} \int_{|\zeta - z| < \epsilon} (f(z) + \mathcal{O}(|\zeta - z|))\, d\lambda(\zeta) = 2\pi i f(z) + \mathcal{O}(\epsilon).$$

Since $\zeta \to (\zeta - z)^{-1}$ is locally integrable, the theorem follows from (1.6) when ϵ tends to zero. □

1.7 Some Simple but Important Consequences.
(a) By Proposition 1.4 and Cauchy's formula (with $f \equiv 1$), we get

$$\int_{|\zeta| = \epsilon} \zeta^{-n} d\zeta = \begin{cases} 2\pi i & \text{if } n = 1 \\ 0 & \text{if } n \neq 1. \end{cases}$$

(b) If the curve Γ starts at a and ends up at b, then, by Proposition 1.4,

$$\int_\Gamma d\zeta = [\zeta]_a^b = b - a.$$

(c) If $\phi \in C_0^1(\mathbb{C})$, then (by Cauchy's formula)

$$\phi(z) = -\frac{1}{\pi}\int \frac{(\partial\phi/\partial\bar\zeta)(\zeta)d\lambda(\zeta)}{\zeta - z} = \frac{\partial}{\partial\bar z}\left(-\frac{1}{\pi}\int \frac{\phi(\zeta)d\lambda(\zeta)}{\zeta - z}\right), \qquad (1.7)$$

where the second equality is obtained by making the linear change of variables $\zeta \mapsto \zeta + z$ in the last integral and differentiating under the integral sign. Hence $(\partial/\partial\bar\zeta)1/\pi\zeta = \delta_0$ (the Dirac measure) in the distribution sense; this is equivalent to $\Delta \log|\zeta|^2 = 4\pi\delta_0$ since $(\partial/\partial\zeta)\log|\zeta|^2 = 1/\zeta$ (even in the distribution sense).

(d) From Cauchy's formula it follows that analytic functions have the *mean value property:*

$$f(z) = \frac{1}{2\pi}\int_0^{2\pi} f(z + re^{it})dt.$$

To see this, one simply makes the substitution $\zeta = z + re^{it}, 0 \le t < 2\pi$ (so that $d\zeta = ire^{it}dt = i(\zeta - z)dt$) in the formula

$$f(z) = \frac{1}{2\pi i}\int_{|\zeta - z| = r} \frac{f(\zeta)d\zeta}{\zeta - z}.$$

(e) If f is analytic and we differentiate under the integral sign in Cauchy's formula, we find that f is C^∞, $f^{(m)}$ is analytic, and

$$f^{(m)}(z) = \frac{m!}{2\pi i}\int_{\partial\omega} \frac{f(\zeta)d\zeta}{(\zeta - z)^{m+1}}, \qquad m = 0, 1, 2, \dots . \qquad (1.8)$$

Thus in particular we have that $A(\Omega) \subset C^\infty(\Omega)$ for any Ω.

1.8 Proposition. *If $K \subset \omega \subset\subset \Omega$, then there are constants $C_m = C_{m,\omega,K}$ such that for all $f \in A(\Omega)$,*

$$\sup_K |f^{(m)}| \le C_m\|f\|_{L^1(\omega)}.$$

Proof. Take $\phi \in C_0^\infty(\omega)$, $0 \le \phi \le 1$, such that $\phi \equiv 1$ in a neighborhood of K. Let $\delta = d(K, \{z \in \omega;\ \phi(z) \ne 1\})$. Since $f\phi \in C_0^\infty(\mathbb{C})$, we get by (1.7)

$$f(z) = (f\phi)(z) = -\frac{1}{\pi}\int \frac{\partial\phi}{\partial\bar\zeta}\frac{f(\zeta)d\lambda(\zeta)}{\zeta - z}, \qquad z \in K. \qquad (1.9)$$

Notice that the integration in this integral is performed only over the strip $\{\zeta;\ 0 < \phi < 1\} \subset\subset \omega \setminus K$. Hence for z in a neighborhood of K we can differentiate the integral and obtain

$$f^{(m)}(z) = -\frac{m!}{\pi}\int \frac{\partial\phi}{\partial\bar\zeta}\frac{f(\zeta)d\lambda(\zeta)}{(\zeta - z)^{m+1}}, \qquad z \in K.$$

In this integral, $|\zeta - z| \geq \delta$, and therefore we get the estimate

$$\sup_K |f^{(m)}| \leq \frac{m!}{\pi \delta^{m+1}} \sup \left| \frac{\partial \phi}{\partial \bar{\zeta}} \right| \int_\omega |f(\zeta)| d\lambda(\zeta).$$

$\square$

The formula (1.9) is a variant of Cauchy's formula where the curve is replaced by a strip. The usual Cauchy formula cannot be used in the preceding proof (since we want L^1-estimates), nor in the next one (since it deals with functions only defined a.e.). However, even in some other situations it is more convenient to use (1.9) rather than the usual Cauchy formula, as will be apparent in what follows.

1.9 Proposition (Weyl's Lemma). *Suppose that* $f \in L^1_{\text{loc}}(\Omega)$ *and* $\partial f / \partial \bar{z} = 0$ *weakly, i.e.,*

$$\int f \partial \phi / \partial \bar{z} = 0, \qquad \phi \in C_0^\infty(\Omega). \tag{1.10}$$

Then there is a $g \in A(\Omega)$ *such that* $f = g$ *a.e.*

Thus, in particular, if $f \in C^0(\Omega)$ and (1.10) holds, then f is analytic. An analogous result is also true (with essentially the same proof) for $f \in \mathcal{D}'(\Omega)$ (the space of distributions on Ω). Clearly, any $f \in A(\Omega)$ satisfies (1.10).

Proof. Take $\omega \subset\subset \Omega$ and $\phi \in C_0^\infty(\Omega)$ such that $\phi \equiv 1$ in a neighborhood of $\bar{\omega}$, and let

$$g(z) = -\frac{1}{\pi} \int \frac{(\partial \phi / \partial \bar{\zeta}) f(\zeta)}{\zeta - z} d\lambda(\zeta), \qquad z \in \omega.$$

If f is analytic, then (1.9) says that $g = f$ in ω; we are going to show that (1.10) actually implies that $f = g$ a.e. in ω. Since $g(z)$ is analytic in ω and $\omega \subset\subset \Omega$ is arbitrary, our proof is then complete. To do this, take $\psi \in C_0^\infty(\omega)$. By Fubini's theorem

$$\int g(z) \psi(z) = -\int_\zeta \frac{\partial \phi}{\partial \bar{\zeta}}(\zeta) \left(-\frac{1}{\pi} \int_z \frac{\psi(z)}{z - \zeta} \right) f(\zeta)$$

$$= -\int_\zeta \frac{\partial}{\partial \bar{\zeta}} \left(\phi(\zeta) \left(-\frac{1}{\pi} \int_z \frac{\psi(z)}{z - \zeta} \right) \right) f(\zeta) + \int_\zeta \phi \frac{\partial}{\partial \bar{\zeta}} \left(-\frac{1}{\pi} \int_z \frac{\psi(z)}{z - \zeta} \right) f(\zeta).$$

The first of these integrals vanishes by the assumption on f, since

$$\phi \left(-\frac{1}{\pi} \int_z \frac{\psi(z)}{z - \zeta} \right)$$

is in $C_0^\infty(\Omega)$ (why?). According to (1.7), the second integral is

$$= \int \phi(\zeta) \psi(\zeta) f(\zeta) = \int \psi(\zeta) f(\zeta)$$

since supp $\psi \subset \{\phi = 1\}$, and thus we have showed that $\int g\psi = \int f\psi$ for all $\psi \in C_0^\infty(\omega)$. This implies that $f = g$ a.e. in ω. $\square$

We say that $f_j \to f$ u.c. (uniformly on compact sets) if for each $K \subset \Omega$, $f_j \to f$ uniformly on K.

1.10 Proposition. *Suppose that* $f_j \in A(\Omega)$.
a) *If* $f_j \to f$ *u.c., then* $f \in A(\Omega)$ *and* $f_j^{(m)} \to f^{(m)}$ *u.c. for each* m.
b) *If* $f_j \to f$ *in* $L_{loc}^1(\Omega)$, *then there is a* $g \in A(\Omega)$ *such that* $f = g$ *a.e. and* $f_j^{(m)} \to g^{(m)}$ *u.c. for each* m.

Proof. The hypothesis in (a) implies the one in (b). To prove (b), observe that f satisfies (1.10) and therefore $f = g$ a.e. for some $g \in A(\Omega)$. Then apply Proposition 1.8 to $f_j - g$. $\square$

One also can easily verify Proposition 1.10 directly from Proposition 1.8 by locally representing f_k by formula (1.9) and taking limits.

1.11 Morera's Theorem. *If* $f \in C(\Omega)$ *and* $\int_{\partial\Delta} f\,dz = 0$ *for all triangles* $\Delta \subset \Omega$, *then* $f \in A(\Omega)$.

Proof. Since analyticity is a local property, we may assume that Ω is convex. Fix a point $a \in \Omega$ and let $F(z) = \int_{[a,z]} f(\zeta)d\zeta$, where $[a,z]$ is the straight line from a to z. By assumption, $\int_{[a,z]} + \int_{[z,z+w]} = \int_{[a,z+w]}$, giving

$$F(z+w) - F(z) = wf(z) + \int_{[z,z+w]} \big(f(\zeta) - f(z)\big)d\zeta = wf(z) + o(|w|)$$

by 1.7 (b) and (1.4). Thus, F is differentiable in z, $\partial F/\partial z = f(z)$, and $\partial F/\partial \bar{z} = 0$; therefore, F is in $C^1(\Omega)$ and hence analytic. Therefore, cf. 1.7 (e), $f = F'$ is analytic. $\square$

Proof of Goursat's Theorem. Assume that Ω is convex, take an arbitrary closed triangle Δ_0 in Ω, and let $I = \int_{\partial\Delta_0} f\,dz$. Divide the triangle into four new triangles Δ^j by inscribing a triangle in Δ_0 with corners at the midpoints of the edges of Δ_0. Then $I = \sum_1^4 \int_{\partial\Delta^j} f\,dz$, which implies that $|\int_{\partial\Delta^j} f\,dz| \geq \frac{1}{4}I$ for some Δ^j. Denote it by Δ_1. By repeating this procedure we get a sequence Δ_k such that $\Delta_k \supset \Delta_{k+1}$ and

$$\left| \int_{\partial\Delta_k} f\,dz \right| \geq 4^{-k}I. \tag{1.11}$$

Choose a point $a \in \cap\Delta_k$. Given $\epsilon > 0$ there is a $\delta > 0$ such that

$$|f(z) - f(a) - (z-a)f'(a)| < \epsilon|z - a|$$

if $|z - a| < \delta$. Take n so large that $\Delta_n \subset \{|z - a| < \delta\}$. Then

$$\left| \int_{\partial \Delta_n} f \, dz \right| = \left| \int_{\partial \Delta_n} (f(z) - f(a) - (z-a)f'(a)) \, dz \right|$$

$$\leq \epsilon \int_{\partial \Delta_n} |z - a| |dz| \leq \epsilon |\partial \Delta_n|^2 = \epsilon 4^{-n} |\partial \Delta_0|^2 \,,$$

which together with (1.11) implies that $I \leq \epsilon |\partial \Delta_0|^2$. Thus, $I = 0$ and the desired conclusion follows from Morera's theorem.   □

1.12 The Maximum (Modulus) Principle. *If Ω is connected, $f \in A(\Omega)$, and $|f(a)| = \sup_\Omega |f|$ for some $a \in \Omega$, then f is constant.*

Proof. From the mean value property, cf. 1.7 (d), we get

$$|f(a)| \leq \frac{1}{2\pi} \int_0^{2\pi} |f(a + re^{it})| \, dt \leq M = \sup_\Omega |f|,$$

with equality in the last inequality if and only if $|f| = M$ on the circle $|\zeta - a| = r$. Hence, if $|f(a)| = M$, then $|f| \equiv M$ in a neighborhood of a. Therefore, the set $A = \{z \in \Omega; |f(z)| = M\}$ is open and closed in Ω. Since $A \neq \emptyset$, $A = \Omega$ and therefore $|f|$, and hence f, is constant.   □

One can replace this closed-open argument, and others in what follows, by a more conceptual "chain of circles" argument.

By the uniqueness theorem (Theorem 2.6 below) it follows that f is constant as long as $|f|$ has a local maximum.

1.13 Corollary. *If $f \in C(\overline{\Omega}) \cap A(\Omega)$ and $\overline{\Omega}$ is compact, then $\sup_\Omega |f| = \sup_{\partial \Omega} |f|$.*

Proof. Take a point $a \in \overline{\Omega}$ such that $|f(a)| = \sup_\Omega |f|$. If $a \in \partial \Omega$, the conclusion is obvious. Otherwise, by the maximum principle, f must be constant in the component of Ω that contains a, but then the supremum also is attained on the boundary.   □

Thus, $|f|$ always attains its supremum on the boundary if Ω is bounded. For unbounded domains this is not true in general. Consider, e.g., $f(z) = \exp(-iz)$ in the upper half-plane. The reader might hope for a bounded counterexample, but that is not possible, cf. Exercise 41.

§2. Power Series Expansions and Residues

For the local study of analytic functions, power series expansions are an important tool, which we are now going to exploit. Suppose that $c_n \in \mathbb{C}$

and $\limsup |c_n|^{\frac{1}{n}} = 1/R$, where $0 < R \leq \infty$. Then the power series

$$f(z) = \sum_{n=0}^{\infty} c_n z^n \qquad (2.1)$$

converges uniformly in each domain $D(0, r)$ (the disk with center 0 and radius r), $r < R$, and diverges if $|z| > R$ (exercise!).

2.1 Proposition. *Suppose that f is given by (2.1). Then f is analytic in $D(0, R)$, $f'(z) = \sum_{n=1}^{\infty} n c_n z^{n-1}$ in $D(0, R)$, and*

$$n! c_n = f^{(n)}(0) = \frac{n!}{2\pi i} \int_{|\zeta| = r} \frac{f(\zeta) d\zeta}{\zeta^{n+1}}, \qquad r < R. \qquad (2.2)$$

Conversely, if $f \in A(D(0, R))$ and c_n are given by (2.2), then $\sum_{n=0}^{\infty} c_n z^n$ converges to f u.c. in $D(0, R)$.

Proof. Let $F_N(z) = \sum_0^N c_n z^n$. Then F_N is analytic in $D(0, R)$, and, since F_N converges u.c., $\lim F_N = f$ is analytic in $D(0, R)$. From Proposition 1.10 we get that $F_N' \to f'$ u.c. and hence $f'(z) = \sum_{n=1}^{\infty} n c_n z^{n-1}$ in $D(0, R)$. By induction it follows that $f^{(n)}(0) = n! c_n$. The second equality in (2.2) is just (1.8).

For the converse, suppose that $|z| < r < R$. Then

$$\frac{1}{\zeta - z} = \sum_0^{\infty} \frac{z^n}{\zeta^{n+1}}$$

converges uniformly for $|\zeta| = r$ and hence

$$f(z) = \sum_{n=0}^{\infty} \frac{1}{2\pi i} \int_{|\zeta| = r} \frac{f(\zeta) d\zeta}{\zeta^{n+1}} z^n = \sum_{n=0}^{\infty} c_n z^n.$$

$\square$

Since

$$|c_n| = \frac{1}{2\pi} \left| \int_{|\zeta| = r} \frac{f(\zeta) d\zeta}{\zeta^{n+1}} \right| \leq r^{-n} \sup_{|\zeta| = r} |f(\zeta)|$$

for each $r < R$, we get

2.2 Cauchy's Estimate. *If $f \in A(D(0, R))$ and $|f| \leq M$, then*

$$|f^{(m)}(0)| = m! |c_m| \leq m! M R^{-m}.$$

2.3 Liouville's Theorem. *If f is an entire function (i.e., analytic in the entire plane) and*

$$|f(z)| \leq C \left(1 + |z|^N \right),$$

then f is a polynomial of degree at most N. In particular, if f is bounded, then it is constant.

Proof. Suppose that $f(z) = \sum_0^\infty c_n z^n$. By Cauchy's estimate,

$$|c_m| \le C(1 + R^N)R^{-m}$$

for all $R > 0$, and consequently $c_m = 0$ if $m > N$.   □

2.4 The Fundamental Theorem of Algebra. *If the polynomial $p(z) = c_N z^N + c_{N-1} z^{N-1} + \cdots + c_0$ has no zeros in $\mathbb{C}$, then it is constant, i.e., $c_j = 0$ for $j > 0$.*

Proof. If p is nonvanishing in $\mathbb{C}$, then $1/p$ is an entire function. Moreover, $1/p(z) = \mathcal{O}(|z|^{-N})$ when $z \to \infty$, and hence is constant by Liouville's theorem.   □

2.5 Proposition. *If $f \in A(D(a,r))$ and f is not identically zero, then for some integer m, $f(z) = (z-a)^m h(z)$, where $h \ne 0$ in a neighborhood of a.*

One then says that $f(z)$ has a *zero of order* or *multiplicity m at a.*

Proof. Since f can be expanded in a power series,

$$f(z) = \sum_{j=m}^\infty c_j(z-a)^j = (z-a)^m \sum_{j=0}^\infty c_{j+m}(z-a)^j = (z-a)^m h(z),$$

where $c_m \ne 0$. Then $h(a) \ne 0$ and hence $h \ne 0$ in a neighborhood of a by continuity.   □

2.6 The Uniqueness Theorem. *If Ω is connected, $f \in A(\Omega)$, and the set $Z(f) = \{z \in \Omega; \ f(z) = 0\}$ has a limit point in Ω, then $f \equiv 0$ in Ω.*

Proof. Let A be the set of limit points of $Z(f)$. Then A is closed in Ω (why?). If $a \in A$, then a is an interior point of A according to Proposition 2.5. Hence A is open. Since Ω is connected and $A \ne \emptyset$, $A = \Omega$.   □

2.7 Remarks. The assumption about connectedness is necessary. If $\Omega = \{z; \ |z| < 1 \ \text{ or } \ |z| > 2\}$, then the function f that is 0 for $|z| < 1$ and 1 for $|z| > 2$ is in $A(\Omega)$ and its zero set has limit points in Ω even though f is not identically zero itself. Observe that if Ω is connected, $f, g \in A(\Omega)$, and $f = g$ on a set with a limit point in Ω, then $f \equiv g$ in Ω. For instance, if the usual trigonometric functions are extended to complex z in the obvious way, $\sin z = (\exp iz - \exp(-iz))/2i$ and so on, then the usual identities, such as $\sin 2z = 2 \sin z \cos z$, hold since they hold for real z.

If $a \in \Omega$ and f is analytic in $\Omega \backslash a$, we say that f has an *isolated singularity* at a. We now shall determine the possible nature of an isolated singularity. Without loss of generality we may assume of course that $a = 0$.

2.8 Proposition. *If f has an isolated singularity at 0 and is bounded in a punctured neighborhood, then the singularity is removable, i.e., f can be defined at 0 so that it is analytic in a neighborhood.*

Proof. Define $h(z)$ as $zf(z)$ for $z \neq 0$ and $h(0) = 0$. Then h is continuous and by Morera's theorem (how?) it is indeed analytic in Ω. Hence, near 0, $h(z) = c_1 z + c_2 z^2 + \ldots$, and therefore $f(z) = c_1 + c_2 z + \ldots$. Thus, f becomes analytic if we let $f(0) = c_1$. □

Notice that the corresponding statement for smooth functions is false; consider, e.g., $z \mapsto \sin(1/|z|)$.

2.9 Proposition. *If f is analytic in $\Omega = \{0 < |z| < R\}$, then f has a Laurent series expansion*

$$f(z) = \sum_{-\infty}^{\infty} c_n z^n$$

that converges u.c. in Ω. The coefficients c_n are given by

$$c_n = \frac{1}{2\pi i} \int_{|\zeta|=r} \frac{f(\zeta)d\zeta}{\zeta^{n+1}}, \quad n \in \mathbb{Z}, \quad 0 < r < R. \tag{2.3}$$

Sketch of Proof. If $0 < \epsilon < |z| < R - \epsilon$, then

$$f(z) = \frac{1}{2\pi i} \int_{|\zeta|=R-\epsilon} \frac{f(\zeta)d\zeta}{\zeta - z} - \frac{1}{2\pi i} \int_{|\zeta|=\epsilon} \frac{f(\zeta)d\zeta}{\zeta - z}.$$

One then proceeds as in the proof of Proposition 2.1 (exercise!). □

More generally: if $f \in A(\{0 < r < |z| < R\})$, then it can be representated by a Laurent series $f(z) = \sum_{-\infty}^{\infty} c_n z^n$ that converges u.c., and the representation is unique since the coefficients must satisfy (2.3).

Definition. Assume that $f(z)$ has an isolated singularity at 0 and Laurent series expansion $\sum_{-\infty}^{\infty} c_n z^n$. If $c_n = 0$ for $n < 0$, the singularity is *removable* (cf. Proposition 2.8). If $c_n = 0$ for $n < -N$ and $c_{-N} \neq 0$, then f has a *pole of order* or *multiplicity* N. Otherwise, f is said to have an *essential singularity* at the origin.

If $f(z)$ has a pole at 0 of order N, then $f(z) - \sum_{-N}^{-1} c_k z^k$ has a removable singularity at 0. The sum is called the *principal part* of $f(z)$ at 0. Moreover,

$z^N f(z)$ has a removable singularity at 0, so $f(z) = z^{-N} h(z)$, where $h(z)$ is analytic and nonvanishing near 0. Hence $|f(z)| \to \infty$ when $z \to 0$, e.g., the image of $0 < |z| < r$ is contained in the complement of any given disk $D(0, R)$ if r is sufficiently small. However, if f has an essential singularity, then it can omit at most one single value in each punctured neighborhood of 0. This is the so-called big Picard theorem, which is proved in Ch. 2. Right now we restrict ourselves to the following much weaker and simpler result.

2.10 Proposition. *Suppose that f is analytic in $\Omega = \{0 < |z| < r\}$ and has an essential singularity at 0. Then its image is dense in $\mathbb{C}$.*

Proof. If the image is not dense, then there is a $w \in \mathbb{C}$ and a neighborhood W of w such that $f(\{0 < |z| < r\}) \cap W = \emptyset$. Then $1/(f(z) - w)$ is bounded, and hence it has a removable singularity. Therefore, $1/(f(z) - w) = z^N h(z)$, where $h(0) \neq 0$, and thus $f(z) = w + 1/z^N h(z)$, i.e., f has a pole of order N at the origin. $\qquad \square$

2.11 Examples.
(a) $\exp(1/z) = \sum_0^\infty z^{-n}/n!$, and therefore $\exp(1/z)$ has an essential singularity at the origin and omits the value zero. By Picard's theorem it therefore attains all other values in each domain $\{0 < |z| < \epsilon\}$. This also can be verified easily by a direct computation (exercise!).
(b) Suppose that $f(z)$ has an isolated singularity at 0 and that $f(z) = \mathcal{O}(|z|^m)$ for some integer m. By Propositions 2.5 and 2.8 there is an integer $N \geq m$ such that $f(z) = z^N h(z)$ and $h(z)$ is analytic and non-vanishing near 0. If $N \geq 0$, then f has a zero of order N; and if $N < 0$, it has a pole of order $-N$.

Exercise 2. Show that the big Picard theorem implies the little Picard theorem: *Each entire nonconstant function attains all values with one possible exception.*

Definition. Suppose that f has an isolated singularity at the point a. For small enough $r > 0$, the number

$$\mathrm{Res}(f, a) = \frac{1}{2\pi i} \int_{|\zeta| = r} f(a + \zeta) d\zeta \tag{2.4}$$

is independent of r and is called the *residue* of f at a; actually, it is just the coefficient c_{-1} in the Laurent expansion of f at a.

If f has a pole of order m at a, then the residue can be computed by the formula (check!)

$$\mathrm{Res}(f, a) = \frac{1}{(m-1)!} \frac{d^{m-1}}{dz^{m-1}}\Big|_a (z - a)^m f(z);$$

but in concrete situations it is often simpler (and, in case of an essential singularity, necessary) to derive c_{-1} in some more direct way.

2.12 Examples. The residue of $\cot z = \cos z / \sin z$ at $z = m\pi$ is 1 since $\lim_{z \to m\pi} (z - m\pi) \cot z = 1$; cf. Remark 2.7.

The function $f(z) = z^2 / \sin(1/z)$ is analytic in $\{0 < |z| < 1/\pi\}$, has an essential singularity at 0, and the residue is $7/360$ since

$$\frac{1}{\sin w} = \frac{1}{w} + \frac{1}{6} w + \frac{7}{360} w^3 + \cdots$$

and hence

$$z^2 \frac{1}{\sin \frac{1}{z}} = z^2 \left(z + \frac{1}{6} \frac{1}{z} + \frac{7}{360} \frac{1}{z^3} + \cdots \right) = z^3 + \frac{1}{6} z + \frac{7}{360} \frac{1}{z} + \cdots.$$

Make this calculation rigorous!

If $f \in C^1(\overline{\omega} \setminus \{a_1, \ldots, a_m\}) \cap A(\overline{\omega} \setminus \{a_1, \ldots, a_m\})$, where $a_1, \ldots, a_m$ are m points in ω, then

$$\int_{\partial \omega} f d\zeta = 2\pi i \sum_{j=1}^{m} \text{Res}(f, a_j).$$

In fact, if ϵ_j is chosen such that $\overline{D(a_j, \epsilon_j)}$ is contained in Ω and does not contain any other a_k, then by Cauchy's theorem (how?)

$$\int_{\partial \omega} f d\zeta = \sum_{j=1}^{m} \int_{|\zeta - a_j| = \epsilon_j} f d\zeta.$$

Definition. A function f is *meromorphic* in Ω if there is a sequence of points $a_k \in \Omega$ that has no limit point in Ω, and if f is analytic in $\Omega \setminus \{a_k\}$ and has a pole at each a_k.

Thus, f is meromorphic in Ω if and only if locally either f or $1/f$ is analytic. Also note that if f is meromorphic in a connected set Ω and its zero set has a limit point in Ω, then $f \equiv 0$.

2.13 Proposition. *Suppose that f is meromorphic in Ω, $\omega \subset \Omega$ and f has no poles or zeros on $\partial \omega$. Let N_f and P_f be the numbers of zeros and poles, respectively (counted with multiplicities). Then*

$$\frac{1}{2\pi i} \int_{\partial \omega} \frac{f'(\zeta)}{f(\zeta)} d\zeta = N_f - P_f.$$

Observe that f has only a finite number of zeros and poles in ω.

Proof. Since $f'(\zeta)/f(\zeta)$ is analytic outside the poles and zeros of f, the integral is unchanged if $\partial \omega$ is replaced by a union of circles, one around each pole or zero a of f. Suppose that $f(z) = (z - a)^m h(z)$, where h is analytic and nonvanishing in $D(a, 2\epsilon)$. Then

$$\frac{f'(z)}{f(z)} = \frac{h'(z)}{h(z)} + m\frac{1}{z - a},$$

and hence

$$\int_{|\zeta - a| = \epsilon} \frac{f'(\zeta)}{f(\zeta)} d\zeta = m 2\pi i.$$

From the above one gets the desired conclusion. □

2.14 Rouché's Theorem. *Assume that $f_t(z): [0, 1] \times \Omega \to \mathbb{C}$ is continuous, $f_t \in A(\Omega)$ for each fixed $t \in [0, 1]$, and also that $f_t'(z): [0, 1] \times \Omega \to \mathbb{C}$ is continuous. If, moreover, f_t is nonvanishing on $\partial\omega$ ($\omega \subset\subset \Omega$), then f_0 and f_1 have the same number of zeros in ω.*

Thus, if we change f_0 continuously, the number of zeros in ω cannot change unless some zero arrives or disappears over the boundary $\partial\omega$. If, for instance, $f, g \in A(\Omega)$ and $|g| < |f|$ on $\partial\omega$, then f and $f - g$ have the same number of zeros in ω; we then simply apply the theorem to $f_t = f - tg$.

Proof. In view of Proposition 2.13,

$$\frac{1}{2\pi i} \int_{\partial\omega} \frac{f_t'(\zeta)}{f_t(\zeta)} d\zeta$$

is integer valued for each t, but since it depends continuously on t, it then must be constant. □

§3. Global Cauchy Theorems

We already have seen that if $\Gamma = \partial\omega$, then

$$\frac{1}{2\pi i} \int_\Gamma \frac{d\zeta}{\zeta - z} = \begin{cases} 1 & \text{if } z \in \omega \quad \text{(by Cauchy's formula)} \\ 0 & \text{if } z \notin \overline{\omega} \quad \text{(by Cauchy's theorem)}. \end{cases}$$

If $\Gamma_1, \ldots, \Gamma_N$ are (piecewise C^1) closed curves with given orientations, we can define the formal sum $\Gamma = \sum_1^N \Gamma_k$. Such a sum is called a cycle. If ϕ is a differential form (1-form), then we let $\int_\Gamma \phi = \sum_1^N \int_{\Gamma_k} \phi$. Note that $\partial\omega$ can be interpreted as a sum of curves with given orientations, i.e., a cycle, but the notion of "cycle" also includes other cases, e.g., passing through a given curve more than one lap.

Now consider a cycle Γ and a point $z \notin \Gamma$. The *index of Γ with respect to z*, $\text{Ind}_\Gamma(z)$, is defined by

$$\text{Ind}_\Gamma(z) = \frac{1}{2\pi i} \int_\Gamma \frac{d\zeta}{\zeta - z}.$$

For all "reasonable" Γ it is clear from above that $\text{Ind}_\Gamma(z)$ is always an integer. This also follows from the next argument, which reveals the geometrical interpretation of $\text{Ind}_\Gamma(z)$ as the winding number. However, we first need a simple lemma.

3.1 Lemma (Local Existence of Logarithm). *If V is a convex domain, $f \in A(V)$ and $f \neq 0$ in V, then there is a (not unique) $h \in A(V)$ such that $\exp h = f$ and $h' = f'/f$.*

It is natural to set $h = \log f$.

Proof. As in the proof of Morera's theorem, we can find a $g \in A(V)$ such that $g' = f'/f$. Thus, $(f\exp(-g))' \equiv 0$ and hence $f\exp(-g) \equiv C$. Take $h = g + a$, where the number a is chosen such that $\exp a = C$. $\qquad\square$

If m is an integer, there is a $g \in A(V)$ such that $g^m = f$. To see this, simply let $g = \exp(h/m)$.

Exercise 3. Show that $\log z = \log|z| + i\arg z$, where the second log denotes the usual real-valued natural logarithm, and arg stands for some branch of the argument function.

Suppose that Γ is a closed curve (the case when Γ is a cycle then follows) with a parametrization $t \to \Gamma(t)$, $0 \leq t \leq 1$. If $0 \neq \Gamma$, there is (exercise!) a sequence of disks $D_0, D_1, \ldots, D_m$ not containing 0 and such that the curve segment $\Gamma_k = \Gamma|_{[t_k, t_{k+1}]}$ is contained in D_k, where $0 = t_0 < t_1 < \cdots < t_m = 1$. We can successively choose branches $\log_k z$ of the logarithm on D_k so that $\log_k z = \log_{k+1} z$ on $D_k \cap D_{k+1}$. Then

$$\int_\Gamma \frac{d\zeta}{\zeta} = \sum_{k=0}^{m-1} \int_{\Gamma_k} \frac{d\zeta}{\zeta} = \sum_{k=0}^{m-1} \left(\log_{k+1} \Gamma(t_{k+1}) - \log_k \Gamma(t_k)\right)$$

$$= i(\arg_m \Gamma(0) - \arg_0 \Gamma(0)),$$

if $\log_k z = \log|z| + i\arg_k z$, and therefore $\text{Ind}_\Gamma(0)$ is the number of laps that Γ runs around the origin. The same holds for $\text{Ind}_\Gamma(z)$.

3.2 Example. The intuitive meaning of index is illustrated by Figure 1.

Exercise 4. Show that Ind_Γ is constant on each component of $\mathbb{C} \setminus \Gamma$ and vanishes on the unbounded one. Hint: Notice that it is continuous in $\mathbb{C} \setminus \Gamma$ and $\mathcal{O}(1/|z|)$ when $|z| \to \infty$.

Suppose that f is meromorphic and has no poles or zeros on the closed curve $\gamma(t)$. Then $\Gamma(t) = f \circ \gamma(t)$ is a closed curve and

$$\text{Ind}_\Gamma(0) = \frac{1}{2\pi i} \int_\gamma \frac{f'(\zeta)}{f(\zeta)} d\zeta.$$

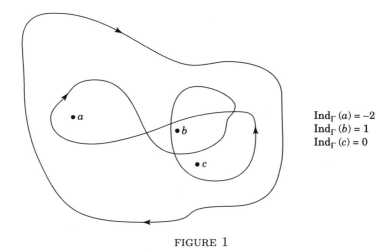

$$\mathrm{Ind}_\Gamma\,(a) = -2$$
$$\mathrm{Ind}_\Gamma\,(b) = 1$$
$$\mathrm{Ind}_\Gamma\,(c) = 0$$

FIGURE 1

Hence, Proposition 2.13 implies that the number of zeros minus the number of poles equals the number of times that $f \circ \gamma(t)$ turns around the origin, if $\gamma(t)$ is a parametrization of $\partial \omega$. This sometimes is called *the principle of argument*. We leave it as an exercise to supply a similar geometrical interpretation of Rouché's theorem.

Definition. We say that a cycle Γ in Ω is *null-homologous* with respect to Ω if $\mathrm{Ind}_\Gamma(a) = 0$ for all $a \notin \Omega$. Two cycles Γ_1 and Γ_2 are *homologous* if $\Gamma_1 - \Gamma_2$ is null-homologous.

Notice that if $\omega \subset\subset \Omega$, then $\partial \omega$ is null-homologous in Ω.

3.3 Cauchy's (Homology) Theorem and Formula. *Suppose that $f \in A(\Omega)$ and Γ is a null-homologous cycle in Ω. Then*

$$\int_\Gamma f(\zeta)d\zeta = 0 \qquad (a)$$

and

$$f(z)\,\mathrm{Ind}_\Gamma(z) = \frac{1}{2\pi i}\int_\Gamma \frac{f(\zeta)d\zeta}{\zeta - z}, \qquad z \in \Omega \setminus \Gamma. \qquad (b)$$

It follows that $\int_{\Gamma_1} f(\zeta)d\zeta = \int_{\Gamma_2} f(\zeta)d\zeta$ if Γ_1 and Γ_2 are homologous. Note that (b) follows from (a), since $\zeta \mapsto (f(\zeta) - f(z))/(\zeta - z)$ is analytic in Ω.

Proof of (a). Let H be the union of Γ and the support of Ind_Γ. As Ind_Γ is locally constant in $\mathbb{C} \setminus \Gamma$ (cf. Exercise 4 above) it must vanish on each component that intersects $\mathbb{C}\setminus\Omega$. However, Ind_Γ vanishes on the unbounded component, and hence H must be a compact set in Ω. Now take $\phi \in C_0^\infty(\Omega)$ such that $\phi \equiv 1$ in a neighborhood of H. Then

$$f(z) = \phi(z)f(z) = -\frac{1}{\pi} \int \frac{\partial \phi}{\partial \bar{\zeta}} \frac{f(\zeta)d\lambda(\zeta)}{\zeta - z}, \qquad z \in H,$$

and by Fubini's theorem (check!)

$$\frac{1}{2\pi i} \int_\Gamma f(z)dz = \frac{1}{\pi} \int \frac{\partial \phi}{\partial \bar{\zeta}} f(\zeta) \operatorname{Ind}_\Gamma(\zeta)d\lambda(\zeta) = 0$$

since $\operatorname{Ind}_\Gamma$ is supported on H and $\partial \phi/\partial \bar{\zeta} \equiv 0$ there. $\square$

3.4 Remark. The preceding proof can be interpreted in the sense of distributions in the following way. Let the distribution (measure) u be defined by $u(\phi) = (i/2) \int_\Gamma \phi d\zeta$ for $\phi \in C_0^\infty$. Then u has compact support in Ω and $(\partial/\partial \bar{\zeta}) \operatorname{Ind}_\Gamma = u$, cf. 1.7 (c). The assumption on Γ implies as before that $\operatorname{Ind}_\Gamma$ is compactly supported in Ω. Thus,

$$\int_\Gamma f d\zeta = u(f) = \frac{\partial \operatorname{Ind}_\Gamma}{\partial \bar{\zeta}}(f) = -\operatorname{Ind}_\Gamma\left(\frac{\partial f}{\partial \bar{\zeta}}\right) = 0.$$

3.5 Remark. Note that $f(z)dz$ is a closed 1-form, i.e., $d(fdz) = 0$ in Ω. By Stokes' theorem (about homology) $\int_\Gamma fdz = 0$ if Γ is null-homologous in Ω. However, to obtain Theorem 3.3 one then has to verify the equivalence of the usual (topological) notion of "null-homologous" and the definition used above, which is natural in this context. In one direction this immediately follows from Stokes' theorem. Conversely, if Γ is null-homologous (in our sense), then the union of components V_k of $\mathbb{C}\backslash\Gamma$ where $\operatorname{Ind}_\Gamma$ is nonvanishing is contained in Ω, and Γ is the boundary of the 2-chain $\sum_k \alpha_k V_k$ where α_k is the value of $\operatorname{Ind}_\Gamma$ on V_k; cf. Example 3.2.

Suppose that γ_0 and γ_1 are two continuous closed curves in Ω, i.e., continuous mappings $[0,1] \to \Omega$ such that $\gamma_j(0) = \gamma_j(1)$. They are *homotopic* if there is a continuous mapping $H: [0,1] \times [0,1] \to \Omega$ such that $H(s,0) = \gamma_0(s)$, $H(s,1) = \gamma_1(s)$, and $H(0,t) = H(1,t)$ for $t \in [0,1]$. Intuitively this means that the curve γ_0 can be continuously deformed to γ_1 within Ω, and vice versa.

Definition. If Ω is connected and any closed curve γ in Ω is homotopic to a point (i.e., to a constant mapping $[0,1] \to \Omega$), then Ω is said to be *simply connected*.

For instance, any convex domain is simply connected; take $H(s,t) = (1-t)\gamma(s) + ta$, where a is any point in Ω.

3.6 Cauchy's (Homotopy) Theorem. *If γ_0 and γ_1 are homotopic closed (piecewise C^1-) curves in Ω and $f \in A(\Omega)$, then $\int_{\gamma_0} f dz = \int_{\gamma_1} f dz$. In particular, $\int_\gamma f dz = 0$ if γ is null-homotopic (homotopic to a point).*

This immediately follows from Cauchy's homology theorem and

3.7 Proposition. *If γ_0 and γ_1 are homotopic closed curves, then they are homologous.*

However, it is just as easy to prove the homotopy theorem directly. Also note that it immediately implies Proposition 3.6.

Sketch of Proof of Cauchy's Homotopy Theorem. Since $H: [0,1]^2 \to \Omega$ is continuous, there is a positive distance 2ϵ from Ω^c to the image of H. One problem is that $s \mapsto \gamma_t(s) = H(s,t)$ is not necessarily piecewise C^1. However, for some large integer n, $|H(s,t) - H(s',t')| < \epsilon$ if $|s-s'| + |t-t'| < 1/n$. Then the rectangle with corners at $H(\frac{i+1}{n}, \frac{k+1}{n})$, $H(\frac{i}{n}, \frac{k}{n})$, $H(\frac{i+1}{n}, \frac{k}{n})$, and $H(\frac{i}{n}, \frac{k+1}{n})$ is contained in Ω. Hence the integral of f over its boundary vanishes. When $k = 0$, instead of the straight line between $H(\frac{i}{n},0)$ and $H(\frac{i+1}{n},0)$, one takes the curve $s \mapsto H(s,0)$, $\frac{i}{n} \leq s \leq \frac{i+1}{n}$, and analogously when $k = n-1$. Let $\tilde{\gamma}_k$ be the polygon with corners $H(\frac{i}{n}, \frac{k}{n})$, $0 \leq i \leq n$. Then we have that $\int_{\tilde{\gamma}_k} f dz = \int_{\tilde{\gamma}_{k+1}} f dz$ (why?). Since we also have that $\int_{\gamma_0} f dz = \int_{\tilde{\gamma}_1}$ and $\int_{\gamma_1} f dz = \int_{\tilde{\gamma}_{n-1}} f dz$, it follows that $\int_{\gamma_0} f dz = \int_{\gamma_1} f dz$. □

3.8 Remark. If a closed curve γ is null-homotopic in Ω, then it is null-homologous by Proposition 3.7. However, the converse is not true in general. Let $\Omega = \mathbb{C} \setminus \{a,b\}$ and choose a point $0 \in \Omega$. Let α be a closed curve in Ω that starts and ends at 0 and surrounds a so that $\text{Ind}_\alpha(a) = 1$ and $\text{Ind}_\alpha(b) = 0$, and let β be a corresponding curve for the point b. Let $\gamma = \alpha\beta^{-1}\alpha^{-1}\beta$ denote the closed curve obtained by first running through β, then through α backwards, then through β backwards, and finally through α. Clearly $\text{Ind}_\gamma(a) = \text{Ind}_\gamma(b) = 0$, and therefore γ is null-homologous by definition. However, at least intuitively, it is perhaps clear that it is not null-homotopic in Ω (a rigorous proof is harder). This reflects the topological fact that the first fundamental group is not commutative. In Figure 2 a curve is drawn that is homotopic to γ.

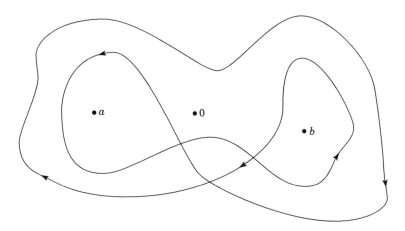

FIGURE 2

Supplementary Exercises

Exercise 5. Suppose that f is C^m in a neighborhood of the origin. Show that

$$f(z) = f(0) + \sum_{k=1}^{m} \frac{1}{k!} \sum_{j=0}^{k} \binom{k}{j} z^{k-j}\bar{z}^{j} \frac{\partial^k f}{\partial z^{k-j}\partial\bar{z}^{j}}\bigg|_0 + o(|z|^m).$$

Exercise 6. Show that an open set Ω in $\mathbb{C}$ is connected if and only if it is pathwise connected (i.e., each pair of points can be connected by a (piecewise C^1) curve in Ω).

Exercise 7. Try to show a variant of Morera's theorem where the triangles are replaced by circles. (Hint: First assume that f is C^1.)

Exercise 8. Assume that f and g are entire functions and that $|f(z)| \leq |g(z)|$ for all z. What can be said about f and g?

Exercise 9. Find all of the zeros of $\sin z$ and $\cos z$.

Exercise 10. Assume that f is analytic in the annulus $\Omega = \{r < |z| < R\}$. Show that f has a Laurent series expansion $f(z) = \sum_{-\infty}^{\infty} c_n z^n$ that converges u.c. in Ω.

Exercise 11. Assume that f has an isolated singularity at 0. Show that
(a) if $f(z) = o(1/|z|)$, then the singularity is removable.
(b) if $f = o(1/|z|^{-(1+N)})$, then f has a pole of order at most N.
(c) if $f(z) = o(|z|^N)$, $N \geq 0$, then the singularity is removable and f has a zero of order at least $N + 1$.

Exercise 12. Compute $\int_{-\infty}^{\infty} (1 + x^2)^{-1} \exp(ixt)dx$ for $t \in \mathbb{R}$.

Exercise 13. Compute $\int_0^\infty (1+x^n)^{-1}dx$, $n = 2, 3, \ldots$. Integrate over an appropriate "piece of cake"!

Exercise 14. Compute $\int_{-\infty}^\infty \frac{\sin x}{x} e^{ixt} dx$ for $t \in \mathbb{R}$.

Exercise 15. Compute $\int_{-\infty}^\infty \exp(-ixt)\exp(-x^2/2)dx$. Note that $t^2 - 2ixt - x^2 = -(x+it)^2$.

Exercise 16. Compute

$$h(z) = -\frac{1}{\pi} \int_{|\zeta|<1} (\zeta - z)^{-1} d\lambda(\zeta)$$

for various $z \in \mathbb{C}$,
(a) using polar coordinates and solving the "θ-integral" by the calculus of residues.
(b) by first computing $(\partial/\partial\bar{z})h(z)$ outside the unit circle and noting that at least $h(z)$ is continuous over the circle.
(c) by expanding the integrand in appropriate power series.

Exercise 17. Compute $\sum_1^\infty (1 + k^2)^{-1}$ by integrating $g(z) = \pi(1 + z^2)^{-1} \cot \pi z$ over the curves Γ_n, which are the rectangles with corners at $(n+1/2)(\pm 1, \pm i)$. Also compute $\sum_1^\infty k^{-2}$.

Exercise 18. Suppose that $f_n \in A(\Omega)$, $|f_n| \le C$ and that $f(z) = \lim f_n(z)$ exists for all $z \in \Omega$. Show that $f_n \to f$ u.c.

Exercise 19. Suppose that $f_n \in A(\Omega)$ and that $f_n \to f$ in $L^p_{\text{loc}}(\Omega)$ for some $p \ge 1$. Show that there is a $g \in A(\Omega)$ such that $f = g$ a.e. and $f_n \to g$ u.c.

Exercise 20. Suppose that $\overline{D(a,r)} \subset \Omega$ and that $\phi, f \in A(\Omega)$. Suppose that f does not vanish on the circle $|z - a| = r$. Determine

$$(2\pi i)^{-1} \int_{|z-a|=r} \frac{f'(z)}{f(z)} \phi(z)dz.$$

Exercise 21. Let $f \in A(\Omega)$, and let γ_0 and γ_1 be two simply closed (piecewise C^1-) curves in Ω. Suppose that there is a homotopy $H(s,t): [0,1]^2 \to \Omega$ from γ_0 and γ_1 such that f has no zero on its image. Show that f has the same number of zeros inside γ_0 and γ_1.

Exercise 22. Give a geometrical interpretation of Rouché's theorem. Also formulate and prove a variant where the functions f_t are allowed to be meromorphic.

Exercise 23. Suppose that $f \in A(\Omega)$, $f(0) = 0$ but $f \not\equiv 0$.
(a) Show that $f(U)$ contains a neighborhood of 0. Hint: For some $\epsilon > 0$, $f \ne 0$ on the circle $|z| = \epsilon$. Consider then

$$\frac{1}{2\pi i} \int_{|\zeta|=\epsilon} \frac{f'(\zeta)d\zeta}{f(\zeta) - w}$$

for w near 0.

(b) Suppose in addition that $f'(0) \neq 0$. Show that f has a local inverse $g(w)$ near 0.

(c) Show that $g(w)$ is in fact analytic. Hint: For w near 0,

$$g(w) = \frac{1}{2\pi i} \int_{|\zeta|=\epsilon} \frac{f'(\zeta)\zeta d\zeta}{f(\zeta) - w}.$$

Exercise 24. Suppose that f and g are nonvanishing analytic functions in U such that $(f'/f)(1/n) = (g'/g)(1/n)$, $n = 1, 2, 3, \dots$. What is the relation between f and g?

Exercise 25. Suppose that Ω is connected, $f_n \in A(\Omega)$ are nonvanishing in Ω, and that $f_n \to f$ u.c. Show that if f is not identically zero, then f is also nonvanishing in Ω.

Exercise 26. Suppose that Ω is connected, $f_n \in A(\Omega)$, and $f_n(\Omega) \subset \Omega'$ for some open Ω'. Show that if $f_n \to f$ u.c., then either f is constant or $f(\Omega) \subset \Omega'$.

Exercise 27. Suppose that $f \in A(U) \cap C(\overline{U})$ and $|f(z)| < 1$ when $|z| = 1$. How many solutions does the equation $f(z) = z$ have in U?

Exercise 28. Suppose that $f \in A(U) \cap C(\overline{U})$, $|f(z)| > 1$ when $|z| = 1$ and $f(0) = 1$. Must f have a zero in U?

Exercise 29. How many zeros does $f(z) = z^5 + 4z^2 + 2z + 1$ have in the unit disk?

Exercise 30. Suppose that f is analytic and $f(z) = \sum_0^m a_k z^k + o(|z|^m)$. Determine a_k.

Exercise 31. Suppose that $f \in A(\Omega)$ and $a_0, \dots, a_m \in \omega \subset\subset \Omega$. Show that

$$p(z) = f(z) - \frac{1}{2\pi i} \int_{\partial\omega} \left(\prod_0^m \frac{z - a_j}{\zeta - a_j} \right) \frac{f(\zeta)d\zeta}{\zeta - z}, \qquad z \in \omega,$$

is the unique polynomial of degree at most m that interpolates f at the points a_j, i.e., such that $f(a_j) = p(a_j)$, $j = 0, \dots, m$.

Exercise 32. Suppose that $f \in C^1(\omega) \cap L^1(\omega)$. Show that

$$u(z) = -\frac{1}{\pi} \int_\omega \frac{f(\zeta)d\lambda(\zeta)}{\zeta - z}, \qquad z \in \omega$$

is a solution to the equation $\partial u/\partial \bar{z} = f$ in ω.

Exercise 33. Supply the details in the following proof that $\mathrm{Ind}_\gamma(0)$ is an integer: If $\gamma \colon [0, 1] \to \mathbb{C}$ is C^1 and closed, $0 \notin \gamma$, and $h(t) = \int_0^t \gamma'(t)/\gamma(t)dt$, then $h'(t) = \gamma'(t)/\gamma(t)$. Hence, $\gamma(t)\exp(-h(t))$ is constant. Therefore, $\exp h(t) = \gamma(t)/\gamma(0)$, and hence $\exp h(1) = 1$.

Exercise 34. Suppose that $f \in A(\Omega)$ and $\omega \subset\subset \Omega$. Show that if $\log |f| \leq \operatorname{Re} g$ on $\partial \omega$ for some $g \in A(\Omega)$, then $\log |f| \leq \operatorname{Re} g$ in ω. Hint: Consider $f \exp(-g)$.

Exercise 35. Let $\Omega = \{x + iy; \ |x| < \pi/2\}$. Show that $f(z) = \exp(\exp iz)$ is unbounded in Ω but $\sup_{\partial \Omega} |f| = 1$.

Exercises 36–38 exemplify the Phragmén–Lindelöf method, which permits certain extensions of the maximum principle to unbounded domains. As an application, examples of complex interpolation then are given in Exercises 39 and 40.

Exercise 36. Suppose that $\Omega = \{x + iy; \ |x| < \pi/2\}$, $f \in A(\Omega) \cap C(\overline{\Omega})$, $|f(z)| \leq \exp(A \exp \alpha |y|)$ for some $\alpha < 1$, $A < \infty$, and that $|f| \leq 1$ on $\partial \Omega$. Show that $|f| \leq 1$ in Ω. Hint:
(a) Let $h_\epsilon(z) = \exp(-2\epsilon \cos \beta z)$, where $\alpha < \beta < 1$. Show that $|h_\epsilon| < 1$ on $\overline{\Omega}$.
(b) Show that

$$|f(z) h_\epsilon(z)| \leq \exp \left(A e^{\alpha |y|} - \epsilon \delta (e^{\beta y} + e^{-\beta y}) \right)$$

on $\overline{\Omega}$ for $\delta = \cos \beta \pi/2$.
(c) Show by the maximum principle that $|f h_\epsilon| \leq 1$ on some large rectangle with corners at $\pm \pi/2 \pm iy_0$.
(d) Conclude that $|f| \leq 1$ in Ω.

Exercise 37. Let $\Omega = \{z; \ |\arg z| < \alpha\}$, $f \in A(\Omega) \cap C(\overline{\Omega})$. Find a boundedness assumption on f which implies that $|f| \leq 1$ in Ω if $\sup_{\partial \Omega} |f| \leq 1$.

Exercise 38. Suppose that $\Omega = \{x + iy; \ a < x < b\}$, $f \in A(\Omega) \cap C(\overline{\Omega})$, and that $|f| \leq C$ in Ω. Let $M(x) = \sup_{y \in \mathbb{R}} |f(x + iy)|$. Show that

$$M^{b-a}(x) \leq M(a)^{b-x} M(b)^{x-a}, \quad a < x < b.$$

Hint:
(a) First suppose that $M(a) = M(b) = 1$ and verify the statement in this case.
(b) Let $g(z) = M(a)^{\frac{b-z}{b-a}} M(b)^{\frac{z-a}{b-a}}$ and apply (a) to $f(z)/g(z)$.

Exercise 39. Suppose that μ and λ are (reasonable) positive measures on two measurable spaces, and suppose that T is a linear operator $L^1(\mu) \to L^\infty(\lambda)$, and $L^2(\mu) \to L^2(\lambda)$ such that $\|Tf\|_\infty \leq M\|f\|_1$ and $\|Tf\|_2 \leq \|f\|_2$. Then $T: L^p(\mu) \to L^q(\lambda)$ for $1 < p < 2$ and $\|Tf\|_q \leq M^{\frac{2-p}{p}} \|f\|_p$ (if q is the dual index). Hint:
(a) It is enough to show:

$$\left| \int (Tf) g d\lambda \right| \leq M^{\frac{2-p}{p}} \quad \forall \ \text{simple} \ f, g \ \text{with} \ \|f\|_p = \|g\|_p = 1. \quad (*)$$

(A function is simple if it is a finite sum $\sum \alpha_j \chi_{E_j}$, where E_j are measurable and have finite measures.)

(b) To show (*), let $(f, g$ simple)

$$\phi(z) = \int |g|^{zp-1} gT(|f|^{zp-1}f)d\lambda.$$

Show that $\phi(z)$ is an entire function, apply Exercise 38 with $a = 1/2$ and $b = 1$, and draw the conclusion that (*) holds.

Exercise 40. Let $\hat{f}(n) = (1/2\pi) \int_0^{2\pi} f(t)e^{-int}dt$ for $f \in L^p(T)$. Show the Hausdorff–Young inequality

$$\left(\sum_{-\infty}^{\infty} |\hat{f}(n)|^q\right)^{1/q} \le \left(\frac{1}{2\pi}\int_0^{2\pi} |f(t)|^p dt\right)^{1/p}, \qquad 1 < p < 2.$$

Exercise 41. Show: If $\Omega \subsetneq \mathbb{C}$ is arbitrary, $f \in A(\Omega) \cap C(\overline{\Omega})$, $|f| \le 1$ on $\partial\Omega$, and $|f| \le M$ in Ω for some M, then $|f| \le 1$ in Ω. Hint:
a) Show that one may assume that $U \cap \Omega = \emptyset$.
b) Take $a \in \Omega$ and apply the maximum principle on $f^n(z)/z$.

Notes

Most of the material in this chapter is classical. For historical remarks about the notion of analytic function, see [Hi].

Notice that the homology variant of Cauchy's theorem (Theorem 3.3) follows from Cauchy's formula and hence in turn from Green's formula in a trivial case.

Exercises 12–17 exemplify the calculus of residues. For hints and further examples, see an elementary book on complex analysis.

The formula in Exercise 31 is called Hermite's formula, and it is fundamental in complex approximation theory.

The technique used in Exercise 39 is called *complex interpolation*. More about this can be found in [S-W].

2

Properties of Analytic Mappings

§1. Conformal Mappings

We now will concentrate on the aspect of analytic functions as mappings from domains in $\mathbb{R}^2$ into $\mathbb{R}^2$. For an arbitrary C^1 mapping the relation

$$f(a+z) - f(a) = \frac{\partial f}{\partial z}\bigg|_a z + \frac{\partial f}{\partial \bar{z}}\bigg|_a \bar{z} + o(|z|)$$

means that the derivative $Df|_a$ of f at a is the real linear mapping

$$z \mapsto \frac{\partial f}{\partial z}\bigg|_a z + \frac{\partial f}{\partial \bar{z}}\bigg|_a \bar{z}. \tag{1.1}$$

In particular, if $(\partial f/\partial \bar{z})|_a = 0$, then $Df|_a$ is just multiplication by the complex number $f'(a) = (\partial f/\partial z)|_a$ and hence nonsingular if (and only if) $f'(a) \neq 0$. The determinant of $z \mapsto f'(a)z$ is equal to $|f'(a)|^2$ (exercise!).

1.1 Proposition. *Suppose that $f \in A(\Omega)$ and $f'(a) \neq 0$. Then there are neighborhoods V and W of a and $f(a)$, respectively, such that f maps V bijectively onto W. Moreover, the inverse g is in $A(W)$ and $g'(f(z)) = 1/f'(z)$.*

Proof. By the inverse function theorem there are open sets V and W such that f maps V bijectively onto W and the inverse $g \in C^1(W)$. Thus, $g(f(z)) = z$ in V, and differentiation of this equality with respect to $\partial/\partial z$ and $\partial/\partial \bar{z}$ yields the remaining statements. $\qquad\square$

A proof without any reference to the inverse function theorem is outlined in Exercise 23 in Ch. 1.

Now we are going to show that $f(\Omega)$ is open if $f \in A(\Omega)$ and f is nonconstant (on each component of Ω). We first consider the function $\pi_m(z) = z^m$, $m > 0$. If $0 \in \pi_m(\Omega)$, then 0 is an interior point in $\pi_m(\Omega)$ since $D(0, r^m) = \pi_m(D(0, r)) \subset \pi_m(\Omega)$ for some $r > 0$, and any other $w \in \pi_m(\Omega)$ is an interior point by Proposition 1.1. A general analytic f can be represented as a composition, as the following result illustrates.

1.2 Theorem. *Suppose that Ω is connected and $f \in A(\Omega)$ is nonconstant. Let m be the order of the zero of $f - f(a)$ at $z = a \in \Omega$. Then there is a neighborhood V of a, $\phi \in A(V)$ and $r > 0$ such that*
(a) *$f(z) = f(a) + (\phi(z))^m$ in V.*
(b) *$\phi' \neq 0$ in V and $\phi: V \to D(0, r)$ is a bijection.*

Thus, $f - f(a) = \tau_m \circ \phi$ in V, and hence f is an m-to-one mapping from $V \setminus \{a\}$ to $\{0 < |w - f(a)| < r^m\}$. In particular, $f(a)$ is an interior point in $f(\Omega)$.

Proof of Theorem 1.2. We may assume that Ω is convex and that $f(z) \neq f(a)$ in $\Omega \setminus \{a\}$. Then $f(z) - f(a) = (z - a)^m g(z)$ in Ω where $g \neq 0$. According to Lemma 3.1 in Ch. 1, there is a function h such that $h^m = g$. If we take $\phi = (z - a)h(z)$, then (a) holds. Since $\phi(a) = 0$ and $\phi'(a) \neq 0$, Proposition 1.1 provides a neighborhood V of a such that (b) also holds. $\square$

Suppose that f is a mapping from an open set Ω in the plane. If there is an $e^{i\phi} \in T$ such that
$$\lim_{r \searrow 0} \frac{f(a + re^{i\theta}) - f(a)}{|f(a + re^{i\theta}) - f(a)|} = e^{i\theta} e^{i\phi} = e^{i(\theta + \phi)}$$
for all θ, we say that f *preserves angles* at a. This means that two half-lines from a meeting at an angle θ are mapped onto two curves meeting at the same angle at $f(a)$, and that the orientation is preserved.

Definition. If f preserves angles at each point, it is called a *conformal mapping*.

If f is differentiable at a and $df|_a \neq 0$ (i.e. $Df|_a \neq 0$), then f preserves angles at a if and only if the mapping (1.1) does, and this in turn happens if and only if $(\partial f/\partial \bar{z})|_a = 0$ (exercise!). Notice, however, that for example $z \mapsto z|z|$ preserves angles at 0 even though the derivative vanishes. Hence any $f \in A(\Omega)$ with a nonvanishing derivative is a conformal mapping. Conversely, if f is a conformal C^1 mapping, it follows that $\partial f/\partial \bar{z} = 0$, where $Df \neq 0$, and therefore $\partial f/\partial \bar{z} \equiv 0$ in Ω, i.e., f is analytic. However, if f is analytic and has a zero of order $m > 1$ at a, then it follows from Theorem 1.2 that f blows up angles m times since $z \mapsto z^m$ does. Thus we have proved the following proposition.

1.3 Proposition. *If $f \in A(\Omega)$ and $f' \neq 0$, then f is a conformal mapping. Conversely, if f is a conformal C^1 mapping in Ω, then f is analytic and $f' \neq 0$.*

Two domains Ω and Ω' in $\mathbb{C}$ are *conformally equivalent* if there is a $\phi \in A(\Omega)$ that maps Ω bijectively onto Ω'. We then find that $\phi^{-1} \in A(\Omega')$ and both ϕ and ϕ^{-1} are conformal mappings. If Ω is conformally equivalent to U, then Ω is homeomorphic to U and hence simply connected. Conversely, we have

1.4 The Riemann Mapping Theorem. *If $\Omega \subset \mathbb{C}$ is simply connected and $\mathbb{C} \setminus \Omega$ is nonempty, then Ω is conformally equivalent to U.*

It follows from Liouville's theorem that $\mathbb{C}$ is not conformally equivalent to U. The Riemann mapping theorem makes it possible to reduce certain problems in simply connected domains into corresponding problems in U. We postpone the proof for a while and begin by studying conformal mappings of U onto itself. To this end we need

1.5 Schwarz' Lemma. *Suppose that $f \in A(U)$, $|f(z)| \leq 1$, and $f(0) = 0$. Then*

$$|f(z)| \leq |z|, \quad z \in U \tag{1.2}$$

and

$$|f'(0)| \leq 1. \tag{1.3}$$

If equality holds in (1.3) or for some z in (1.2), then $f(z) = \lambda z$ for some $\lambda \in T$.

Proof. Set $f_r(z) = f(rz)$ for $r < 1$. Then $f_r(z)/z$ is in $A(U) \cap C(\overline{U})$, and by the maximum principle

$$|f_r(z)/z| \leq \sup_{|\zeta|=1} |f_r(\zeta)/\zeta| \leq 1, \quad z \in U,$$

i.e., $|f_r(z)| \leq |z|$. Letting r tend to 1 we get (1.2) and thereby (1.3). Observe that $g(z) = f(z)/z$ ($g(0) = f'(0)$) is in $A(U)$ and that $|g(z)| \leq 1$. If $|g(z)| = 1$ for some $z \in U$, then $g(z) = \lambda$ for some $\lambda \in T$, again according to the maximum principle.   □

Exercise 1. Suppose that $f \in A(U)$, $|f(z)| \leq 1$, and that $f(z)$ has a zero of order m at 0. Show that $|f(z)| \leq |z|^m$. What happens if there is equality?

A bijective conformal mapping of Ω onto itself is called an *automorphism*. The set of automorphisms of Ω is a group under composition.

1.6 Automorphisms of U. For $\alpha \in U$ we let

$$\phi_\alpha(z) = \frac{z - \alpha}{1 - \bar{\alpha}z}.$$

Then ϕ_α is analytic in the entire plane except for a pole at $1/\bar{\alpha}$, which is outside $\overline{U}$. For $|z| = 1$ we have $|z - \alpha| = |\bar{z} - \bar{\alpha}||z| = |1 - \bar{\alpha}z|$, and therefore T is mapped into itself, and hence by the maximum modulus principle U is mapped into itself (this also follows from the elementary inequality $|z - \alpha| \leq |1 - \bar{\alpha}z|$ for $z \in U$). However, since $\phi_\alpha \circ \phi_{-\alpha}(z) = z$ (check!), ϕ_α actually maps T onto itself and U onto itself bijectively. For any $\lambda \in T$ and $\alpha \in U$ we have that $\psi(z) = \lambda\phi_\alpha(z)$ is an automorphism of U. We claim that every automorphism is of this form. In fact, if $f(z)$ is an automorphism, then $f(\alpha) = 0$ for some $\alpha \in U$, and thus $g = f \circ \phi_{-\alpha}$ is an automorphism such that $g(0) = 0$. Hence, by Schwarz' lemma, $|g(z)| \leq |z|$. However, g^{-1} is also an automorphism, and therefore $|z| \leq |g(z)|$. Hence, $|g(z)| = |z|$ and thus $g(z) = \lambda z$ for some $\lambda \in T$, i.e., $f = \lambda\phi_\alpha$.

Exercise 2. Suppose that $\psi: \Omega \to U$ is a bijective analytic mapping. Show that *any* automorphism of Ω has the form $\psi^{-1}(\lambda\phi_\alpha \circ \psi)$. Also show that if $a, b \in \Omega$, then there is an automorphism such that $g(a) = b$, i.e., the automorphism group is transitive.

Let us now consider the class Σ of analytic functions that map U into U. For a given $\alpha \in U$, we want to optimize $|g'(\alpha)|$. If $g \in \Sigma$, then Schwarz' lemma applies to $f = \phi_{g(\alpha)} \circ g \circ \phi_{-\alpha}$, giving $|f'(0)| \leq 1$, and equality holds if and only if $f(w) = \lambda w$, i.e., $g(z) = \phi_{-g(\alpha)}(\lambda\phi_\alpha(z))$. However,

$$f'(0) = \phi'_{g(\alpha)}(g(\alpha))g'(\alpha)\phi'_{-\alpha}(0) = (1 - |g(\alpha)|^2)^{-1}g'(\alpha)(1 - |\alpha|^2),$$

giving

$$|g'(\alpha)| \leq \frac{1 - |g(\alpha)|^2}{1 - |\alpha|^2}.$$

Summing up, we get: If $g \in \Sigma$, then

$$|g'(\alpha)| \leq \frac{1}{1 - |\alpha|^2},$$

and the maximum is attained if and only if $g(z) = \lambda\phi_\alpha(z)$. In particular, any optimizing g is automatically bijective and $g(\alpha) = 0$.

Exercise 3. Suppose that Ω is conformally equivalent to the unit disk. Let Σ be the set of all $f \in A(\Omega)$ with $|f(z)| < 1$ and fix $\alpha \in \Omega$. Show that there is an $f \in \Sigma$ such that $|f'(\alpha)| = \sup_{g\in\Sigma} |g'(\alpha)|$, and that this f is a bijective mapping of Ω onto U with $f(\alpha) = 0$.

Hence, if there is any analytic bijective mapping of Ω onto U, one can find such a mapping by optimizing $|f'(\alpha)|$. This is the idea behind the proof of the Riemann mapping theorem. However, to carry out the proof we need some further preparations.

Definition. A set $\Phi \subset A(\Omega)$ is called a normal family if any sequence f_n from Φ has a subsequence that converges u.c. in Ω (the limit function is not necessarily in Φ). Sometimes it is convenient to make the definition less restrictive and accept subsequences that tend to ∞ u.c. For our purposes one can adopt either definition.

1.7 Proposition. *If $\Phi \subset A(\Omega)$ and Φ is uniformly bounded on compacts, then it is a normal family.*

Proof. It follows from Proposition 1.8 of Ch. 1 that the gradients of the functions in Φ are uniformly bounded on compacts and hence Φ is equicontinuous on compacts. The conclusion then follows from the Arzela–Ascoli theorem.   □

1.8 Lemma. *If Ω is connected, $f_n \in A(\Omega)$, $f_n \to f$ u.c., all f_n are injective, and f is not constant, then f is also injective.*

Proof. Take $a \in \Omega$ and observe that $g_n(z) = f_n(z) - f_n(a)$ are nonvanishing in $\Omega \setminus \{a\}$ and that $g_n \to f - f(a) \equiv g$ u.c. Since g is nonconstant and $\Omega \setminus \{a\}$ is connected, g is also nonvanishing in $\Omega \setminus \{a\}$ (this follows for instance from the maximum theorem or Rouché's theorem; cf. Exercise 25 of Ch. 1) and hence $g(z) \neq g(a)$ if $z \neq a$.   □

1.9 Proposition. *If Ω is simply connected, then each $f \in A(\Omega)$ has a primitive, and hence each nonvanishing $f \in A(\Omega)$ has a logarithm.*

Proof. Take $\alpha \in \Omega$ and define

$$F(z) = \int_{\Gamma_z} f d\zeta,$$

where Γ_z is a curve from α to z; since Ω is simply connected, $F(z)$ is independent of the choice of Γ_z (why?). As in the proof of Morera's theorem, it follows that $F' = f$. The last claim then follows; cf. Lemma 3.1 in Ch. 1.□

Proof of the Riemann Mapping Theorem. Let Σ be the set of all $f \in A(\Omega)$ that are injective (or *univalent*) and map Ω into U. By assumption there is some $w \in \mathbb{C} \setminus \Omega$, and since Ω is simply connected, there is, by Proposition 1.9, a $\phi \in A(\Omega)$ such that $\phi^2(z) = z - w$. If $\phi(z_1) = \pm\phi(z_2)$, then $z_1 = z_2$, and therefore ϕ is injective and $\phi(z_1) \neq -\phi(z_2)$. Since $\phi(\Omega)$ is open, it contains a disk $D(a, r)$ with $0 < r < |a|$. Thus $D(-a, r) \cap \phi(\Omega) = \emptyset$ and therefore $\psi = r/(\phi + a) \in \Sigma$; hence Σ is nonempty.

Choose $\alpha \in \Omega$. The next step consists in proving that if $\psi \in \Sigma$ and $\psi(\Omega)$ is not the entire U, then there is a $\tilde{\psi} \in \Sigma$ such that

$$|\tilde{\psi}'(\alpha)| > |\psi'(\alpha)|. \tag{1.4}$$

To this end take $\beta \in U \setminus \psi(\Omega)$. Then $\phi_\beta \circ \psi \in \Sigma$ is nonvanishing in Ω, and therefore there is a $g \in A(\Omega)$ such that $g^2 = \phi_\beta \circ \psi \in \Sigma$. However,

g is injective (why?) and therefore $\tilde{\psi} = \phi_{g(\alpha)} \circ g \in \Sigma$. If $\pi(z) = z^2$, then $\psi = \phi_{-\beta} \circ \pi \circ \phi_{-g(\alpha)} \circ \tilde{\psi}$ and therefore $\psi'(\alpha) = (\phi_{-\beta} \circ \pi \circ \phi_{-g(\alpha)})'(0)\tilde{\psi}'(\alpha)$; but since $\phi_{-\beta} \circ \pi \circ \phi_{-g(\alpha)} : U \to U$ is *not* injective, the modulus of its derivative at 0 is less than one, and hence (1.4) follows.

Now only the "soft" part remains. Let $\eta = \sup_{\psi \in \Sigma} |\psi'(\alpha)|$. It follows from the above that $\psi \in \Sigma$ is onto U if $|\psi'(\alpha)| = \eta$ (note that $\eta > 0$ since Σ is nonempty and that $\eta < \infty$ since Σ is uniformly bounded). Choose a sequence ψ_n from Σ such that $|\psi'_n(\alpha)| \to \eta$. By Proposition 1.7, Σ is a normal family, and hence we can extract a subsequence (which we also denote ψ_n) that converges u.c. to some ψ. Since $|\psi'(\alpha)| = \eta > 0$, ψ is non-constant and hence injective by Lemma 1.8. Since $\psi_n(\Omega) \subset U$, $\psi(\Omega) \subset \overline{U}$, but $\psi(\Omega)$ is open and therefore $\psi(\Omega) \subset U$. Thus, we have found a $\psi \in \Sigma$ with optimal derivative, and by the preceding step it must be surjective; hence, the theorem is proved.   □

§2. The Riemann Sphere and Projective Space

We compactify the plane $\mathbb{C}$ by adjoining a point ∞ so that the disks $D(\infty, r) = \{|z| > r\} \cup \{\infty\}$ constitute a basis of neighborhoods at ∞ and denote it $\mathbb{P}$, the one-dimensional (complex) *projective space*. By the mapping

$$\Phi(re^{i\theta}) = (1 + r^2)^{-1}(2r\cos\theta, 2r\sin\theta, r^2 - 1), \quad \Phi(\infty) = (0, 0, 1),$$

we can identify $\mathbb{P}$ with the sphere S^2 in $\mathbb{R}^3$. We leave it as an exercise to verify that Φ is indeed a homeomorphism and to find out its geometrical meaning.

In $D(\infty, 0)$, $w = 1/z$ is continuous. By taking w as a complex coordinate we can define such concepts as analyticity, zero, removable singularity, pole, and essential singularity in this set. This is consistent since in the set $0 < |z| < \infty$, the function $f(z)$ is analytic if and only if $z \mapsto f(1/z)$ is, and so on. Thus, from the point of view of the z coordinate, a function f defined in $D(\infty, r)$ is *analytic* if $z \mapsto f(1/z)$ is analytic in $D(0, 1/r)$, and it has a zero at ∞ if and only if $w \mapsto f(1/w)$ has a zero at $w = 0$ and so on. Since f is meromorphic in $\Omega \subset \mathbb{P}$ if and only if locally either f or $1/f$ is analytic, a meromorphic function in Ω is just an analytic mapping from Ω to $\mathbb{P}$.

2.1 Remark. Topologically, $\mathbb{P}$ is just S^2 via the homeomorphism Φ. Anyway, we prefer the notion $\mathbb{P}$ since it emphazises the complex structure, even though it is often practical to keep in mind the geometrical picture of the sphere, which in this context is referred to as the *Riemann sphere*. Moreover, even though $\mathbb{P}$ is invariant in a certain sense (see Exercise 32), we mostly think of it as an extension of the usual complex plane, so that for instance the points 0, 1, and ∞ have a specified meaning.

2.2 Remark. If $w = \phi(z)$ is a bijective analytic function in Ω, then concepts like analyticity, pole of order m, essential singularity as well as others can be defined just as well with respect to the coordinate w. This immediately follows from the chain rule. However, the value of the residue at a point depends on the particular choice of coordinate.

To clarify at this point, recall that any 1-form can be written uniquely as $f\,dz + g\,d\bar{z}$. This decomposition is invariant under any analytic change of coordinate. The first term is a $(1,0)$-*form* and the other one a $(0,1)$-*form*.

Since integration along a curve is intrinsically defined for 1-forms, it follows from the definition (cf. formula (2.4) of Ch. 1) that the value of the residue at some point α is invariantly defined for the meromorphic $(1,0)$-form $f(z)\,dz$. (In the same way, the value of the derivative of an analytic f at a certain point depends on the choice of coordinate, whereas the form $df = f'(z)\,dz$ is invariantly defined.) For instance, dz is a meromorphic form with residue 0 at infinity, whereas the form dz/z has residue 1 at the origin and -1 at ∞. It is easily verified (exercise!) that if f is a global meromorphic form on $\mathbb{P}$, then the sum of all of its residues is 0.

Since $\mathbb{P}$ is compact, any global analytic function on $\mathbb{P}$ must attain its maximum at some interior point and hence by the maximum principle f is constant, since $\mathbb{P}$ is connected. However, there are nontrivial meromorphic functions on $\mathbb{P}$. By the compactness of $\mathbb{P}$ a meromorphic function f has just a finite number of poles. Let $l(z)$ be the sum of the principal parts at these points. Recall that the principal part at a point $\alpha \in \mathbb{C}$ is the "negative part" of the Laurent expansion with respect to z at this point. If ∞ is a pole, we take instead the principal part with respect to $w = 1/z$. Then $f - l$ is analytic on $\mathbb{P}$ and hence constant. Therefore, $f = l + c$ is rational, and we obtain the partial fraction decomposition

$$f(z) = p(z) + \sum_{k=1}^{m} \sum_{j=1}^{n(k)} \frac{c_{jk}}{(z - a_k)^j},$$

where $n(k)$ is the order of the pole in a_k and p is a polynomial ($p - p(0)$ is the principal part at ∞).

2.3 Example. We now can give a new approach to Louiville's theorem. If $f(z)$ is an entire function that is $\mathcal{O}(|z|^m)$ when $|z| \to \infty$, then it has a pole of order at most m at ∞. Therefore, there is a polynomial $p(z)$ of degree at most m (the principal part at ∞ with respect to $w = 1/z$) such that $f(z) - p(z)$ is analytic at ∞ and hence constant by the maximum principle, so $f(z) = c + p(z)$.

2.4 Example. Let us determine all automorphisms of the entire plane. Certainly any affine mapping $z \mapsto az + b$ is an automorphism of $\mathbb{C}$, and in fact all automorphisms are of this kind. In fact, if f is an automorphism, then it has an isolated singularity at ∞; but since f is injective and open, it follows that there is some punctured neighborhood of ∞ where f avoids the open set $f(U)$. Thus, it follows from Proposition 2.10 in Ch. 1 that the singularity is a pole (or removable). By Louiville's theorem then f is a polynomial and hence linear because of its injectivity.

Let M be the set of all linear fractions,

$$\phi(z) = \frac{az + b}{cz + d},$$

where $ad \neq bc$. Each $\phi \in M$ is an analytic bijective mapping from $\mathbb{P}$ to $\mathbb{P}$, and one easily checks that M is closed under composition and that each $\phi \in M$ has an inverse in M, i.e., M is a group.

Exercise 4. Show that if α, β, γ are three different points on $\mathbb{P}$, there is a unique $\phi \in M$ such that $\phi(\alpha) = 0$, $\phi(\beta) = 1$, and $\phi(\gamma) = \infty$.

Exercise 5. Show that $\phi \in M$ maps "circles" onto "circles," if "circle" means circle or line in $\mathbb{C}$.

2.5 Automorphisms of $\mathbb{P}$. We claim that M is precisely the automorphism group of $\mathbb{P}$. To begin with, assume that $g: \mathbb{P} \to \mathbb{P}$ is an automorphism such that $g(0) = 0$, $g(1) = 1$ and $g(\infty) = \infty$. We already know that g must be rational, and since it has no pole in $\mathbb{C}$, it is a polynomial. Since it is injective, it then must be a linear polynomial, and thus $g(z) = z$ since $g(0) = 0$ and $g(1) = 1$. Now if $f: \mathbb{P} \to \mathbb{P}$ is any automorphism, we can compose it with a linear fraction ϕ such that $g = \phi \circ f$ has 0, 1, and ∞ as fixed points. However, then $\phi \circ f$ is the identity and therefore $f = \phi^{-1} \in M$.

A further discussion about $\mathbb{P}$ and its automorphism group M is outlined in the exercises.

§3. Univalent Functions

Suppose that f is analytic in U, $f(0) = 0$, and $f'(0) = 1$. Then we know from Proposition 1.1 that $f(U)$ contains a neighborhood of the origin. Is there a constant r such that $rU \subset f(U)$ for all such f? Without further restrictions on f the answer is no. For instance, the function $f_\epsilon(z) = \epsilon(\exp(z/\epsilon) - 1)$ satisfies these requirements, but the point $-\epsilon$ is not contained in $f_\epsilon(U)$. However, if we impose the extra condition that f be injective, or *univalent*, then the situation is different. The principal result is Koebe's theorem.

Definition. Let S be the class of injective analytic functions in U with $f(0) = 0$ and $f'(0) = 1$.

Notice that each $f \in S$ is a conformal equivalence from U to its image $f(U)$.

3.1 Theorem (Koebe's Theorem). *If $f \in S$, then*

$$f(U) \supset D(0, 1/4).$$

The main tool in the proof of Koebe's theorem is a result referred to as the area theorem. In the proof we use the following simple observation.

Exercise 6. If ω is a bounded domain with C^1 boundary, then

$$\int_{\partial\omega} \bar{z}\,dz = 2i \int_\omega d\lambda(z).$$

3.2 Theorem (The Area Theorem). *Suppose that*

$$h(z) = \frac{1}{z} + c_0 + c_1 z + c_2 z^2 + c_3 z^3 + \dots$$

is analytic and injective for $0 < |z| < 1$. Then

$$\sum_1^\infty n|c_n|^2 \le 1.$$

Notice that if h is as in the theorem and avoids the value a, then $1/(h(z)-a) \in S$. Conversely, if $f \in S$, then $h = 1/f$ satisfies the hypothesis of the theorem.

Proof. Take $r < 1$. Since h is injective and analytic, it is actually an orientation preserving diffeomorphism from U to a certain subset of $\mathbb{P}$ containing ∞; hence $V_r = \mathbb{P} \setminus h(\overline{D(0,r)})$ is a an open subset of the plane whose boundary is parametrized by $\gamma\colon \theta \mapsto h(re^{i\theta})$ and θ runs from 2π to 0 (sic!). In virtue of the preceding exercise, the area of V_r is

$$A_r = \frac{1}{2i}\int_\gamma \bar{w}\,dw = -\frac{1}{2i}\int_0^{2\pi} \overline{h(re^{i\theta})}h'(re^{i\theta})ire^{i\theta}\,d\theta. \qquad (3.1)$$

A simple calculation gives that

$$\overline{h(re^{i\theta})}h'(re^{i\theta})re^{i\theta} = -\frac{1}{r^2} + (|c_1|^2 + 2|c_2|^2 + 3|c_3|^2 + \dots)r^2 + \dots,$$

where the dots at the end denote terms with nonzero integer powers of $e^{i\theta}$. Since $\int_0^{2\pi} e^{im\theta}\,d\theta = 0$ for $m \ne 0$, (3.1) implies that

$$0 \le A_r = \pi\left(\frac{1}{r^2} - \sum_1^\infty n|c_n|^2 r^2\right).$$

The desired result is obtained by letting r tend to 1.   □

Notice that equality in the area theorem means that the intersection of the sets V_r is a zero set, i.e., h attains almost all values. The theorem implies in particular that $|c_1| \le 1$. If we have $|c_1| = 1$, then $c_k = 0$ for $k > 1$ and hence

$$h(z) = \frac{1}{z} + c_0 + e^{i\theta} z.$$

This function is actually injective (verify!) and maps $\{0 < |z| < 1\}$ onto the complement of the line segment between the points $-2e^{i\theta/2} + c_0$ and $2e^{i\theta/2} + c_0$. (Since h is continuous on $\overline{U}$ and $h(U)$ is open, it is enough to verify that h maps T onto this segment.) In particular, if we choose $c_1 = 1$ and $c_0 = 2$, then $h(z)$ avoids the interval $[0, 4]$ and hence $f = 1/h \in \mathcal{S}$ avoids the set $[1/4, \infty)$, which shows that the constant $1/4$ in Koebe's theorem is the best possible.

3.3 Proposition. *Suppose that $f = z + a_2 z^2 + \ldots \in \mathcal{S}$.*
(a) There is a function $g \in \mathcal{S}$ such that $g^2(z) = f(z^2)$.
(b) $|a_2| \le 2$.

Proof. Since $f(z)$ is injective, $f(z)/z$ is nonvanishing and therefore equal to ϕ^2 for a ϕ with $\phi(0) = 1$. Thus $f(z^2) = (g(z))^2$, where

$$g(z) = z\phi(z^2) = z + \frac{1}{2} a_2 z^3 + \cdots,$$

so to prove (a) we just have to verify that $g(z)$ is injective. However, if $g(z) = g(w)$, then $f(z^2) = f(w^2)$ so that $z = \pm w$; but $g(-z) = -g(z)$ and so $z = -w$ implies that $z = w = 0$. Hence, in any case $z = w$.

For the second statement, letting $g(z)$ be as above we have that

$$\frac{1}{g(z)} = \frac{1}{z}\left(1 - \frac{1}{2} a_2 z^2 + \cdots\right) = \frac{1}{z} - \frac{1}{2} a_2 z + \cdots,$$

and therefore by the area theorem, $|(-1/2)a_2| \le 1$. $\qquad \square$

3.4 Remark. If we have equality in (b), then as noted above $1/g(z) = 1/z + c_0 + e^{i\theta} z$. The relation $f(z^2) = g^2(z)$ then implies that $c_0 = 0$ and thus $f(z) = z/(1 + e^{i\theta} z)^2$. We leave it as an exercise to determine the range of this function. Note that

$$f(z) = e^{-i\theta} \sum_{n=1}^{\infty} (-1)^{n+1} n e^{in\theta} z^n,$$

and hence $|a_n| = n$. In fact, for any $f = z + a_2 z^2 + \cdots \in \mathcal{S}$ it is true that $|a_n| \le n$. For a long time this was known as Bieberbach's conjecture, and it was proved by de Branges in 1984.

We have seen an example of an injective $h(z) = 1/z + c_0 + c_1 z + \ldots$ in U that avoided an interval of length 4. Actually this is optimal, as we have

3.5 Proposition. *If $h(z) = 1/z + c_0 + c_1 z + \ldots$ is injective in U and avoids the values w_1 and w_2, then $|w_1 - w_2| \leq 4$.*

Proof. By assumption, $1/(h(z) - w_j) = z + (w_j - c_0)z^2 + \cdots \in S$, so $|w_j - c_0| \leq 2$ by Proposition 3.3 (b). This implies that $|w_1 - w_2| \leq 4$. □

It is now easy to prove Koebe's theorem.

Proof of Theorem 3.1. If $f \in S$ and avoids w, then $h = 1/f$ avoids $1/w$ and 0, and therefore $|1/w| \leq 4$. Thus, $w \in f(U)$ if $|w| < 1/4$. □

§4 Picard's Theorems

Picard's theorem states that if an entire function omits two distinct values, then it is constant; and, more generally, that if an analytic function omits two distinct values in some punctured neighborhood of an isolated singularity, then the singularity is either a pole or removable. The latter statement is usually called the big Picard theorem. The proof presented here proceeds via a classical result called Schottky's theorem. The main tool is Bloch's theorem, which states that there is an absolute constant ℓ such that if $f \in A(U)$ and $f'(0) = 1$, then $f(U)$ contains a disk with radius ℓ. It should be pointed out that in general this disk will not be centered at $f(0)$; cf. the example at the beginning of the preceding paragraph. However, if one also imposes a boundedness condition, then in fact $f(U)$ contains $D(f(0), r)$ for some fixed r.

4.1 Proposition. *Suppose that $f \in A(U)$ satisfies that $f(0) = 0$ and $f'(0) = 1$. If furthermore $|f| \leq M$, then*

$$f(U) \supset D(0, 1/4M).$$

Notice that if $M = 1$, then $f(U) = U$ by Schwarz' lemma. The proposition implies that

$$f(D(0, R)) \supset D(0, |f'(0)|^2 R^2 / 4M)$$

if f is analytic in $D(0, R)$, $f(0) = 0$, and $|f| \leq M$.

Proof. If $w \notin f(U)$, there is a $h \in A(U)$ with $h(0) = 1$ such that

$$h(z) = (1 - f(z)/w)^{1/2} = 1 - \frac{1}{2w}z + \ldots$$

and $|h(z)|^2 \leq 1 + M/|w|$. For any $h(z) = a_0 + a_1 z + \ldots \in A(U)$ we have

$$\frac{1}{2\pi} \int_0^{2\pi} |h(re^{i\theta})|^2 d\theta = \sum_0^\infty r^{2n} |a_n|^2, \qquad r < 1,$$

and therefore $1 + r^2/4|w|^2 \leq 1 + M/|w|$, i.e., $|w| \geq r^2/4M$. Letting $r \to 1$, we get the desired result. $\qquad \square$

4.2 Theorem (Bloch's Theorem). If $f \in A(U)$ and $f'(0) = 1$, then $f(U)$ contains some disk with radius ℓ, where ℓ is an absolute constant.

If f is analytic in $D(a, R)$, it follows that the image of f contains a disk with radius $|f'(a)|R\ell$.

For an even stronger statement also due to Bloch, see Exercise 31.

Proof. We may assume that f is analytic in some neighborhood of the closure of U since otherwise we can consider $f(rz)/r$ for r near 1. Let

$$w(t) = t \sup_{|z| \leq 1-t} |f'(z)|.$$

Then $w(t)$ is continuous for $0 \leq t \leq 1$, $w(0) = 0$, and $w(1) = 1$, and therefore there is a least $t_0 > 0$ such that $w(t_0) = 1$. Choose a such that $|a| \leq 1 - t_0$ and $|f'(a)| = 1/t_0$. In $D(a, t_0/2)$, $|f'| \leq 2/t_0$ since $D(a, t_0/2) \subset D(0, 1 - t_0/2)$ and in the latter domain $\sup |f'| < 1/(t_0/2) = 2/t_0$. Now

$$g(z) = f(z) - f(a) = \int_a^z f'(\tau) d\tau$$

is analytic in $z \in D(a, t_0/2)$, $|g'(a)| = 1/t_0$, and $|g(z)| \leq t_0/2 \cdot 2/t_0 = 1$, and therefore $g(D(a, t_0/2))$ contains the disk $D(0, 1/16)$, i.e., $f(U)$ contains the disk $D(f(a), 1/16)$. $\qquad \square$

We now can conclude the little Picard theorem. With no loss of generality we may assume that f is an entire function that omits the values 0 and 1. Since $\mathbb{C}$ is simply connected, there is some function $\log f$ in U and furthermore some function

$$g(z) = \log \left[\sqrt{\frac{\log f}{2\pi i}} - \sqrt{\frac{\log f}{2\pi i} - 1} \right].$$

We claim that this g omits the set

$$E = \left\{ \pm \ln(\sqrt{n} - \sqrt{n-1}) + m2\pi i, \quad n \geq 1, \ m = 0, \pm 1, \pm 2, \ldots \right\}.$$

In fact,

$$f = \exp \left(2\pi i \left(\frac{\exp g + \exp(-g)}{2} \right)^2 \right) \qquad (4.1)$$

(let $a = \log f/2\pi i$; then $\exp g = \sqrt{a} - \sqrt{a-1}$ and $\exp(-g) = \sqrt{a} + \sqrt{a-1}$, and thus $\exp g + \exp(-g) = 2\sqrt{a}$). Certainly there is a positive number d such that each disk with radius d intersects E, and hence it follows from Bloch's theorem that $g'(a) = 0$ for any a. Therefore, g is constant and hence so is f.

To obtain the big Picard theorem, we need Schottky's inequality, the proof of which requires a somewhat refined version of the previous argument.

4.3 Proposition (Schottky's Inequality). *There are positive functions* $m(\alpha, \beta, r)$ *and* $M(\alpha, \beta, r)$ *such that if* $f \in A(U)$ *avoids 0 and 1 and* $\alpha \le |f(0)| \le \beta$, *then*

$$m(\alpha, \beta, r) \le |f(z)| \le M(\alpha, \beta, r), \qquad |z| \le r < 1.$$

Proof. Assume that $f \in A(U)$ and avoids the values 0 and 1. Since U is simply connected, there is a function $g \in A(U)$ such that (4.1) holds. Moreover, we can choose g by first determining the value $g(0)$ so that

$$|g(0)| \le C(\alpha, \beta) \qquad \text{if} \quad 0 < \alpha \le |f(0)| \le \beta. \tag{4.2}$$

If $|z| = r < 1$, then the function

$$\phi(\zeta) = \frac{g(z + (1-r)\zeta)}{(1-r)g'(z)}$$

is analytic in the unit disk and $\phi'(0) = 1$. By Bloch's theorem, therefore, its image contains a disk with radius ℓ, i.e., some disk with radius $(1-r)|g'(z)|\ell$ is contained in $g(U)$. However, since g avoids the set E, we must have that

$$|g'(z)| < \frac{d}{\ell(1-r)}.$$

Since

$$g(z) = g(0) + \int_0^z g'(\tau)d\tau,$$

we thus obtain the estimate

$$|g(z)| < |g(0)| + \frac{d|z|}{\ell(1-|z|)} \le C(\alpha, \beta) + \frac{d|z|}{\ell(1-|z|)}. \tag{4.3}$$

From (4.3) and (4.2) we now get that $|f(z)| < M(\alpha, \beta, r)$ if $|z| \le r < 1$. Applying it to $1/f$ we get the lower estimate. □

Notice that one needs both the upper and the lower bound on $|f(0)|$ even to get the estimate $|f(z)| \le M$.

4.4 The Big Picard Theorem. *If f is analytic near a point a and omits two distinct values in some punctured neighborhood of a, then a is a pole or a removable singularity.*

Proof. We have to prove that if f is analytic for $|z| > 1/2$ and avoids 0 and 1, then either f is bounded or has a pole at infinity. If not, the image of f on each set $|z| > R$ is dense according to Proposition 2.10 in Ch. 1, and therefore there are $\lambda_n \to \infty$ such that $1/2 \le |f(\lambda_n)| \le 1$. Set $f_n(z) = f(\lambda_n z)$. Then f_n is analytic in $|z| > 1/2$ and $1/2 \le |f_n(1)| \le 1$, and therefore Schottky's inequality yields that

$$c \le |f_n(z)| \le C, \qquad z \in D(1, 1/4).$$

For $\alpha \in T \cap D(1, 1/4)$ we can repeat the argument and hence get constants c' and C' such that

$$c' \le |f_n(z)| \le C', \qquad z \in D(\alpha, 1/4).$$

After a finite number of steps we have covered the entire unit circle, and hence there are constants k and K such that

$$k \le |f_n(z)| \le K, \qquad z \in T,$$

i.e., $|f| \le K$ on $|z| = \lambda_n$. By the maximum principle

$$|f| \le \max\left(K, \sup_T |f|\right)$$

for $1 \le |z| \le |\lambda_n|$. Hence, f is bounded for $1 \le |z| < \infty$, which is a contradiction. $\square$

Supplementary Exercises

Exercise 7. Show that the maximum modulus principle follows from the open mapping theorem.

Exercise 8. Show that the determinant of the linear mapping $z \mapsto \alpha z$ is $|\alpha|^2$.

Exercise 9. Complete the proof of Proposition 1.3, i.e., show that if $df|_a \ne 0$ and f preserves angles at a, then $(\partial f/\partial \bar{z})|_a = 0$.

Exercise 10. Given $\Omega \subset \mathbb{C}$, find an exhausting sequence of compacts $\ldots K_n \subset K_{n+1} \ldots$, i.e., such that any compact $K \subset \Omega$ is contained in some K_n. Find an increasing sequence of compacts K_n such that $\cup K_n = \Omega$ which is not exhausting.

Exercise 11. Let $\Phi = \left\{f \in A(U); \int_{|\zeta|<r} |f| d\lambda \le \exp(1-r)^{-1}\right\}$. Is Φ a normal family?

Exercise 12. Show that the function Φ from $\mathbb{P}$ (the extended plane) to S^2 is indeed a homeomorphism. Include a drawing.

Exercise 13. Suppose that $f: U \to U$ is analytic and $f(\alpha) = \beta$. Show that

$$\left| \frac{f(z) - \beta}{1 - \bar{\beta}f(z)} \right| \leq \left| \frac{z - \alpha}{1 - \bar{\alpha}z} \right|.$$

Exercise 14. Find all automorphisms of the upper half-plane Π^+.

Exercise 15. Suppose that $f \in A(\Pi^+)$ and $|f| < 1$. How large can $|f'(i)|$ be? Which f are optimal?

Exercise 16. Suppose that $f \in A(U) \cap C(\bar{U})$ and $|f| = 1$ on T. Show that f has at least one zero in U unless f is constant.

Exercise 17. Find all $f \in A(U) \cap C(\bar{U})$ such that $|f| = 1$ on T. Hint: Note that each finite product of functions ϕ_α is of this kind.

Exercise 18. Suppose that $f, g \in A(U)$, $f(0) = g(0)$, $g(U) \subset f(U)$, and that f is injective. Show that $g(D(0, r)) \subset f(D(0, r))$, $0 < r < 1$.

Exercise 19. Let $\Phi = \{f \in A(U); \operatorname{Re} f > 0 \text{ and } f(0) = 1\}$. Show that Φ is a normal family. What happens if one removes the assumption $f(0) = 1$?

Exercise 20. Find a homeomorphism from U to U that has no continuous extension to $\bar{U}$.

Exercise 21. Suppose that Ω is connected and symmetric with respect to the real axis, $f \in A(\Omega)$, and f is real on $\Omega \cap \mathbb{R}$. Show that $\overline{f(z)} = f(\bar{z})$.

Exercise 22. Suppose that f is analytic in a connected neighborhood of T and that $|f| = 1$ on T. Show that $f(z) = 1/\overline{f(1/\bar{z})}$.

Exercise 23. Find all meromorphic functions f in $\mathbb{C}$ such that $|f| = 1$ on T.

Exercise 24. Let $A(1, R) = \{z; 1 < |z| < R\}$. Show that if $A(1, R)$ is conformally equivalent to $A(1, R')$, then $R = R'$ by filling out the details in the following arguments. Let $f: A(1, R) \to A(1, R')$ be analytic and bijective. First show that $|f(z_j)| \to 1$ when $|z_j| \to 1$ and $|f(z_j)| \to R'$ when $|z_j| \to R$, or vice versa. In that case, replace f by R'/f. Then show that

$$\log|f(z)| = \frac{\log R'}{\log R} \log|z|$$

and conclude that $R = R'$. Hint:

$$u(z) = \log|f(z)| - (\log R'/\log R)\log|z|$$

is harmonic in $A(1, R)$ and vanishes on the boundary, and hence it vanishes identically; see Ch. 4.

Exercise 25. Find all automorphisms of $A(1, R)$.

Exercise 26. Show that if $f(z)dz$ is a meromorphic form on $\mathbb{P}$, then the sum of all of its residues is 0.

Exercise 27. Show that if $f \in \mathcal{S}$, then there is a $g \in \mathcal{S}$ such that $g^m(z) = f(z^m)$.

Exercise 28. Suppose that $f(z)$ is an injective meromorphic function in U such that $f(0) = 0$ and $f'(0) = 1$, and suppose that f avoids w_1 and w_2. Show that $|w_1 - w_2|/|w_1||w_2| \leq 4$.

Exercise 29. Determine the image of the function $f(z)$ in Remark 3.4.

Exercise 30. Here is a sketch of an alternative proof of Proposition 4.1. If $f(z) = z + a_2 z^2 + a_3 z^3 + \ldots$, then $|a_k| \leq M$ by Cauchy's estimate. For $|z| = r < 1$,

$$|f(z)| \geq r - Mr^2 - Mr^3 - \ldots = r(1 + M) - \frac{rM}{1 - r} = \phi(r),$$

and $\phi(r)$ attains its maximum at $\rho = 1 - \sqrt{M/(1 + M)}$. However, since (we may assume that) $M \geq 1$,

$$\phi(\rho) = (\sqrt{1 + M} - \sqrt{M})^2 = (\sqrt{1 + M} + \sqrt{M})^{-2} > \frac{1}{(1 + \sqrt{2})^2 M} > \frac{1}{6M}.$$

Thus, if $|z| = r \leq \rho$, then

$$|f(z)| \geq \phi(r) \geq r\frac{\phi(\rho)}{\rho}$$

as $r \mapsto \phi(r)/r$ is decreasing. Hence, $f(z)$ has just one zero in $D(0, \rho)$. However, if $|z| = \rho$ and $|w| < \phi(\rho)$, then $|f(z)| \geq \phi(\rho) > |w|$ and thus by Rouché's theorem, $z \mapsto f(z) - w$ has exactly one zero in $D(0, \rho)$. Thus, $f(D(0, \rho)) \supset D(0, \phi(\rho))$.

Exercise 31. Show the following stronger version of Theorem 4.2: *There exists an absolute constant $b > 0$ such that if $f \in A(\Omega)$ and $f'(0) = 1$, then some open subset of U is mapped bijectively by f onto some disk with radius b.*

The following exercises contain a further discussion about the one-dimensional complex projective space $\mathbb{P}$ and the Riemann sphere.

Exercise 32. Here is the usual formal definition of $\mathbb{P}$: On $\mathbb{C}^2 \setminus \{(0,0)\}$ the points $Z = (z_0, z_1)$ and $W = (w_0, w_1)$ are equivalent if and only if $w = \alpha z$ for some $\alpha \in \mathbb{C} \setminus \{0\}$. The set of equivalence classes is denoted by $\mathbb{P}$ and is made into a topological space by requiring that the natural projection $\pi : \mathbb{C}^2 \setminus (0,0) \to \mathbb{P}$ be continuous and open. Since $\pi|_{\{|Z|=1\}}$ is surjective, $\mathbb{P}$ is compact. In $\pi(\{z_0 \neq 0\})$, one can take $z = z_1/z_0$ as a local complex coordinate. A nonsingular linear transformation on $\mathbb{C}^2$ with matrix

$$G = \begin{pmatrix} d & c \\ b & a \end{pmatrix}$$

gives rise to an automorphism $\psi(z)$ on $\mathbb{P}$ given by the formula

$$\phi(z) = \frac{az + b}{cz + d} \qquad (*)$$

in the coordinate z. It follows that the set M of all automorhisms of $\mathbb{P}$ of the type $(*)$ is a group.

Exercise 33. From this point of view a natural distance or metric on $\mathbb{P}$ is

$$\chi(Z, W) = \frac{|w_0 z_1 - w_1 z_1|}{\sqrt{|z_0|^2 + |z_1|^2}\sqrt{|w_0|^2 + |w_1|^2}},$$

which in the coordinate $z = z_1/z_0$ is

$$\chi(z, w) = \frac{|z - w|}{\sqrt{1 + |z|^2}\sqrt{1 + |w|^2}}, \qquad \chi(z, \infty) = \frac{1}{\sqrt{1 + |z|^2}}.$$

Actually, $\chi(z, w) \geq 0$ and $\chi(z, w) = 0$ if and only if $z = w$ and the triangle inequality holds, i.e., $\chi(z, w) \leq \chi(z, u) + \chi(u, w)$. Moreover, $\chi(z, w) \leq 1$ with equality if and only if $w = -1/\bar{z}$. The metric χ is invariant under the subgroup $\mathcal{U}$ of M, that arises from the group of unitary matrices G, i.e., such that $G^*G = I$. The subgroup $\mathcal{U}$ is (also) transitive, i.e., for any pair of points $z, w \in \mathbb{P}$ there is some $\psi \in \mathcal{U}$ mapping z on w.

Exercise 34. Each "circle" on $\mathbb{P}$ is of the form $\{z;\ \chi(z, w) = r\}$.

Exercise 35. There is a unique normalized measure $d\mu$ on $\mathbb{P}$ that is invariant under $\mathcal{U}$, i.e., such that $\mu(E) = \mu(\psi(E))$ for all (measurable) E and $\phi \in \mathcal{U}$. Since

$$\lim_{h \to 0} \frac{\chi(z + h, z)}{|h|} = \frac{1}{1 + |z|^2},$$

it follows that

$$d\mu(z) = \frac{d\lambda(z)}{\pi (1 + |z|^2)^2}.$$

Exercise 36. Show that if $\mathbb{P}$ is identified with the sphere S^2 via the mapping Φ, then χ corresponds to half the cordial distance between points on S^2. Moreover, show that the measure $d\mu$ corresponds to the normalized surface measure on S^2. Also show that a "circle" on $\mathbb{P}$ corresponds to a circle on S^2.

Notes

The first complete proof of the Riemann mapping theorem is from the end of the nineteenth century. The proof presented here depends heavily on normal families. A more constructive proof was obtained by Koebe.

For the history of the Riemann mapping theorem and its connection to Dirichlet's problem, see [Hi].

Any doubly connected domain (such that the complement has two components) with reasonable boundary is conformally equivalent to an annulus $A(1, R)$ (for a unique R in view of Exercise 24). See [A1] and [A2] for further results about multiply connected domains.

Koebe proved in 1907 that there exists some constant k such that $f(U) \supset D(0, k)$ for each $f \in \mathcal{S}$. The optimal value $k = 1/4$ was determined by Bieberbach in 1916. Bieberbach also made the conjecture that $|a_n| \leq n$ for all n. de Brange's proof of the conjecture is published in *Acta Math.*, vol. 154 (1985).

Picard proved the theorem bearing his name in 1879. The proof presented here is due to Bloch (1924). The result in Exercise 31, as well as Theorem 4.2, is due to Bloch. The best constant in Theorem 4.2 is called Landau's constant L, whereas the best constant in Exercise 31 is called Bloch's constant B. Their exact values still are not determined, but it is known that $0.4330\ldots < B < 1/2 < L < 0.5433\ldots$; see Liu and Minda, *Trans. Amer. Math. Soc.*, vol. 333 (1992), 325–338.

3

Analytic Approximation and Continuation

§1. Approximation with Rationals

In this section we study the possibility of approximating analytic functions with polynomials and, more generally, by rational functions. The main result is Runge's theorem.

1.1 Runge's Theorem. *Suppose that K is compact in $\mathbb{P}$ and that the set $\{a_j\}$ contains one point from each component of $\mathbb{P} \setminus K$. If f is analytic in a neighborhood of K, then, for given $\epsilon > 0$, there is a rational function r having poles only in the set $\{a_j\}$ such that $|f - r| < \epsilon$ on K.*

Of course, one of the points may be ∞. Note that $\mathbb{P} \setminus K$ has at most countably many components.

1.2 Corollary. *If $K \subset \mathbb{C}$ and $\mathbb{C} \setminus K$ is connected, i.e., K has no "holes," then any f that is analytic in a neighborhood of K can be uniformly approximated by polynomials on K.*

Notice that there is no requirement that K be connected.

1.3 Example. There is a sequence of polynomials p_n such that $p_n(z) \to 1$ for all $z \in \overline{U}$ and $p_n(z) \to 0$ for all $z \notin \overline{U}$. In order to see this, let $K_n = \overline{U} \cup \{z;\ 1 + 1/n \leq |z| \leq n, 0 \leq \arg z \leq 2\pi - 1/n\}$. Since the complement of each K_n is connected, Runge's theorem applies to the function f that is 1 in a neighborhood of $\overline{U}$ and 0 in a neighborhood of $K_n \setminus \overline{U}$; therefore, there are polynomials $p_n(z)$ such that $|p_n - 1| < 1/n$ on $\overline{U}$ and $|p_n| < 1/n$ on $K_n \setminus \overline{U}$. The sequence p_n then has the stated properties.

The requirement of (at least) one point from each component of $\mathbb{P} \setminus K$ is necessary. In fact, if a is a point in the component V of $\mathbb{P} \setminus K$, then

$f(z) = (z - a)^{-1}$ is analytic in a neighborhood of K (take $f(z) = z$ if $\infty \in V$). If $f(z)$ could be approximated arbitrarily well on K by a rational function $r(z)$ with no poles in V, then $(z - a)r(z) - 1$ would be near 0 on $\partial V \subset K$ and hence in V by the maximum principle. This leads to a contradiction when $z = a$.

Proof of Runge's Theorem. We may assume that $K \subset \mathbb{C}$. The dual space of $C(K)$ is precisely the space of (complex) measures on K. If $f|_K$ is not in the closure of the space (of restrictions to K) of rational functions having poles in $\{a_j\}$, there is (according to the Hahn–Banach theorem) a measure μ on K that vanishes on these rational functions but such that $\mu(f) \neq 0$.

Thus, if μ is a measure on K that vanishes on rational functions with poles in each $\{a_j\}$, we have to prove that $\mu(f) = 0$. To this end, let

$$h(\zeta) = \int_K \frac{d\mu(z)}{z - \zeta}, \quad \zeta \in \mathbb{P} \setminus K.$$

Let V be a component of $\mathbb{P} \setminus K$ and a_k a point in $V \cap \{a_j\}$. If $a_k \neq \infty$, then

$$\frac{1}{z - \zeta} = \sum_0^\infty \frac{(\zeta - a_k)^n}{(z - a_k)^{n+1}}$$

uniformly for $z \in K$ if $\zeta \in D(a_k, r) \subset\subset V$. Hence, $h = 0$ in $D(a_k, r)$ by the assumption on μ, and since h is analytic in V, it must vanish identically there; analogously if $a_k = \infty$. Hence, $h \equiv 0$ in $\mathbb{P} \setminus K$. Now take $\phi \in C_0^\infty(\mathbb{C})$ such that $\phi \equiv 1$ in a neighborhod of K. Then for $z \in K$,

$$f(z) = f(z)\phi(z) = -\frac{1}{\pi} \int \frac{f(\zeta)\partial\phi/\partial\bar{\zeta}}{\zeta - z} d\lambda(\zeta)$$

and, by Fubini's theorem,

$$\mu(f) = \int_K f d\mu = \frac{1}{\pi} \int f(\zeta)\frac{\partial\phi}{\partial\bar{\zeta}}h(\zeta)d\lambda(\zeta) = 0,$$

since $\partial\phi/\partial\bar{\zeta} = 0$ on K and $h = 0$ outside K.   □

We leave it to the reader to give an interpretation of this proof in the sense of distributions; cf. Remark 3.4 in Ch. 1 and Exercise 18. A constructive proof of Runge's theorem is outlined in Exercise 6.

By means of the following simple lemma we obtain some important corollaries of Runge's theorem.

1.4 Lemma. If $\Omega \subset \mathbb{C}$, then there are compact sets $\ldots K_n \subset \text{int}(K_{n+1}) \subset K_{n+1} \subset \cdots \subset \Omega$ such that each component of $\mathbb{P} \setminus K_n$ intersects $\mathbb{P} \setminus \Omega$ and any $K \subset \Omega$ is contained in some K_n.

The first condition on K_n means that it has no "unnecessary holes."

Sketch of Proof. Take a sequence $\ldots H_n \subset \text{int}(H_{n+1}) \subset H_{n+1} \subset \cdots \subset \Omega$ such that $\cup H_n = \Omega$. Let K_n be the union of H_n and all components of $\mathbb{P} \setminus H_n$ that do not intersect $\mathbb{P} \setminus \Omega$.   □

One also can define K_n directly by $\mathbb{P} \setminus K_n = D(\infty, n) \cup \cup_{a \notin \Omega} D(a, 1/n)$. By the lemma we get the following corollaries:

1.5 Corollary. *If $\Omega \subset \mathbb{P}$ and $f \in A(\Omega)$, there are rational functions r_n with poles outside Ω such that $r_n \to f$ u.c. in Ω.*

1.6 Corollary. *If $\Omega \subset \mathbb{C}$, $\mathbb{P} \setminus \Omega$ is connected and $f \in A(\Omega)$, then there are polynomials p_n such that $p_n \to f$ u.c. in Ω.*

§2. Mittag–Leffler's Theorem and the Inhomogenous Cauchy–Riemann Equation

Recall that a *principal part* at $a \in \mathbb{C}$ is a rational function

$$p(z) = \sum_{j=1}^{n} \frac{c_j}{(z-a)^j}.$$

2.1 Mittag–Leffler's Theorem. *Let a_j be a sequence with no limit point in $\Omega \subset \mathbb{C}$, and let $p_j(z)$ be principal parts at a_j. Then there is a meromorphic f in Ω that has principal part p_j at a_j for each a_j and no other poles.*

Proof. Take K_n as in Lemma 1.4 and let

$$q_n(z) = \sum_{a_j \in K_n} p_j(z).$$

There is only a finite number of a_j in K_n; therefore, each q_n is a rational function and $q_{n+1} - q_n$ is analytic in a neighborhood of K_n. Hence by Runge's theorem there are rational r_n with poles in $\mathbb{P} \setminus \Omega$ such that $|q_{n+1} - q_n - r_n| < 2^{-n}$ on K_n. Define

$$f = q_1 + \sum_{1}^{\infty} (q_{n+1} - q_n - r_n).$$

For fixed N,

$$f = q_N - \sum_{1}^{N-1} r_n + \sum_{N}^{\infty} (q_{n+1} - q_n - r_n),$$

and the last sum converges uniformly on K_N and hence is analytic in int(K_N). Since the r_n have their poles outside Ω, f is meromorphic in int(K_N) and has the prescribed principal parts there (and no others). Since N is arbitrary, the proof is complete. $\qquad\qquad\square$

2.2 Remark. The notion of principal part p of f at the point a depends on the coordinate z. However, in any case, $f - p$ is analytic near a, and therefore the principal parts with respect to any two different coordinates differ by an analytic function. Mittag–Leffler's theorem can be stated in the following invariant way: *Suppose that we have an open cover Ω_j of $\Omega \subset \mathbb{P}$ and meromorphic f_j in Ω_j such that $f_j - f_k$ is analytic in $\Omega_j \cap \Omega_k$. Then there is a global meromorphic function f in Ω such that $f = f_j$ in Ω_j.* It is readily verified that Theorem 2.1 follows from this statement, a proof of which is outlined in Exercise 7.

Recall (cf. (1.7) in Ch. 1) that if $\phi \in C_0^\infty$, then

$$u(z) = -\frac{1}{\pi} \int \frac{\phi(\zeta) d\lambda(\zeta)}{\zeta - z} \qquad (2.1)$$

is a C^∞ solution to $\partial u / \partial \bar{z} = \phi$. By a proof analogous to the preceding proof of Mittag–Leffler's theorem we get an existence theorem for solutions to the inhomogenous Cauchy–Riemann equation.

2.3 Theorem. *If $\Omega \subset \mathbb{C}$ and $f \in C^\infty(\Omega)$, there is a $u \in C^\infty(\Omega)$ such that*

$$\frac{\partial u}{\partial \bar{z}} = f \qquad (2.2)$$

in Ω.

Proof. Take K_n as before and choose $\phi_n \in C_0^\infty(\text{int}(K_{n+1}))$ such that $\phi_n \equiv 1$ in a neighborhood of K_n. Now there are $u_n \in C^\infty(\Omega)$ such that $\partial u_n / \partial \bar{z} = \phi_n f$. Observe that $(\partial / \partial \bar{z})(u_{n+1} - u_n) = 0$ in a neighborhood of K_n. Thus there are rational r_n, having their poles outside Ω, such that $|u_{n+1} - u_n - r_n| < 2^{-n}$ on K_n. Then

$$u = u_1 + \sum_1^\infty (u_{n+1} - u_n - r_n)$$

is a C^∞-function, since on each fixed compact there is only a finite number of terms that are not analytic and the "tail" converges uniformly. Therefore, we may differentiate termwise, and hence

$$\frac{\partial u}{\partial \bar{z}} = \phi_1 f + \sum_1^\infty (\phi_{n+1} - \phi_n) f = f.$$

$\qquad\qquad\square$

2.4 Remark. For a function u let $\bar{\partial}u = (\partial u/\partial\bar{z})d\bar{z}$. Then $\bar{\partial}u$ is the $(0,1)$-part of the form df and hence invariant under analytic changes of coordinate; cf. Remark 2.2 in Ch. 2. If F is a smooth $(0,1)$ form, then the equation $\bar{\partial}u = F$ has an invariant meaning, even at infinity. We then can formulate the preceding theorem as: *If $\Omega \subset \mathbb{P}$ and f is a smooth $(0,1)$-form in Ω, then there is a smooth function u such that $\bar{\partial}u = f$ in Ω.* This follows immediately from Theorem 2.3 except in the case $\Omega = \mathbb{P}$. However, if F is a global $(0,1)$-form on $\mathbb{P}$, we can solve $\bar{\partial}u_j = F$ in $\Omega_0 = D(0,2R)$ and $\Omega_1 = D(\infty, R)$, respectively. Then $u_1 - u_0$ is analytic in the annulus $\{R < |z| < 2R\}$, and therefore (using, e.g., the Laurent series expansion) the function can be written $h_1 - h_0$, where h_j are analytic in Ω_j. Thus $u_0 - h_0 = u_1 - h_1$ in the annulus and, since $\Omega_0 \cup \Omega_1 = \mathbb{P}$, we get a global solution.

Theorem 2.3 is a pure existence theorem. It is clear that if u_0 is a particular solution to (2.2), then any other solution has the form $u = u_0 + h$ where $h \in A(\Omega)$. Often it is important to single out a solution with certain properties. This can be done by L^2 methods (see Ch. 8, where some examples of this technique are exploited in the text and in the exercises).

Another way to get further information about a solution is to use some explicit formula. If $f \in L^1(\Omega)$, then

$$u(z) = -\frac{1}{\pi}\int_\Omega \frac{f(\zeta)d\lambda(\zeta)}{\zeta - z}$$

provides a solution to (2.2). [To see this, let χ be a cut-off function in Ω that is identically 1 in some neighborhood of $K \subset \Omega$ and write $f = \chi f + (1-\chi)f$. This gives rise to two terms: the first one solves (2.2) on K (cf. (2.1)), and the second one is analytic near K.] It is also possible to construct such formulas that will work for a larger class of f:

2.5 Example. It follows that

$$Kf(z) = -\frac{1}{\pi}\int_U \left(\frac{1 - |\zeta|^2}{1 - \bar{\zeta}z}\right)\frac{f(\zeta)d\lambda(\zeta)}{\zeta - z}$$

solves (2.2) if $\int_U(1-|\zeta|^2)|f| < \infty$. In fact, one can just apply the arguments above to the function

$$\zeta \mapsto \left(\frac{1 - |\zeta|^2}{1 - \bar{\zeta}w}\right)f(\zeta)$$

and let $z = w$. In this formula, f is thus allowed to grow somewhat at the boundary. However, this solution also has a certain minimality property.

Let $B(U)$ be the subset of analytic functions in U that belong to $L^2(U)$. From Proposition 1.10 of Ch. 1 it follows that $B(U)$ is a closed subspace of $L^2(U)$. Let

$$Pu(z) = \frac{1}{\pi} \int_U \frac{u(\zeta)d\lambda(\zeta)}{(1 - \bar{\zeta}z)^2}.$$

One can verify (see Exercise 9) that P is a bounded operator from $L^2(U)$ to $B(U)$. Furthermore, one can check that

$$K(\partial u/\partial \bar{z}) = u - Pu \qquad (2.3)$$

for $u \in C^1(\overline{U})$. Since, for instance, the analytic polynomials are dense in $B(U)$ (Exercise 9), (2.3) shows that $Pu = u$ if $u \in B(U)$ is analytic, i.e., P is a projection. Moreover, it is indeed the orthogonal projection $P: L^2(U) \rightarrow B(U)$ since its kernel $(1 - \bar{\zeta}z)^{-2}/\pi$ is hermitian (which in turn implies that P is self-adjoint). The operator P is called the *Bergman projection* in U and $B(U)$ the *Bergman space*. Now, if (2.2) has a solution u in for example $L^2(U) \cap C^1(\overline{U})$, it follows from (2.3) that Kf is the one with minimal norm in $L^2(U)$ since it is orthogonal to $B(U)$.

Further examples of integral representation of solutions will be given in the exercises.

§3. Analytic Continuation

In Ch. 1 we found that the logarithm function could be defined along any curve in the plane that avoids the origin. This is an instance of *analytic continuation*, a notion that we now shall make precise. A *function element* is an ordered pair (f, D), where D is an open disk and $f \in A(D)$. Suppose that $\gamma: [0, 1] \rightarrow \Omega$ is a curve and we are given $f_0 \in A(D(\gamma(0), r_0))$. We say that f_0 (or more precisely $(f_0, D(\gamma(0), r_0))$) *can be continued along* γ if there are numbers $0 = s_0 < s_1 < \cdots < s_n = 1$ and function elements (f_j, D_j) such that $\gamma(0)$ is the center of D_0, $\gamma(1)$ is the center of D_n, $\gamma([s_j, s_{j+1}]) \subset D_j$, and $f_j = f_{j+1}$ in $D_j \cap D_{j+1}$ (which by the assumption is nonempty). Note that if g is a continuation of f along γ, then f is a continuation of g along $-\gamma$.

3.1 Proposition. *Any two continuations of f_0 along γ coincide, i.e., if we have $0 = s_0 < s_1 < \cdots < s_n = 1$, $0 = \sigma_0 < \sigma_1 < \cdots < \sigma_m = 1$, and corresponding function elements (f_j, D_j) and $(g_k, \tilde{D}_k)$, then $f_n = g_m$ in $D_n \cap \tilde{D}_m$.*

Hence, we can talk about *the continuation* of f_0 along γ.

Proof. Assume the contrary. Then there are i and j such that $[s_i, s_{i+1}] \cap [\sigma_j, \sigma_{j+1}] \neq \emptyset$, $f_i \neq g_j$ on $D_i \cap \tilde{D}_j$, and $i+j$ is minimal. We also may assume that $s_i \geq \sigma_j$. Then $i \geq 1$ and $s_i \in [\sigma_j, \sigma_{j+1}]$. Thus, $\gamma(s_i) \in D_{i-1} \cap D_i \cap \tilde{D}_j$.

Since $i + j$ is minimal and $[s_{i-1}, s_i] \cap [\sigma_j, \sigma_{j+1}] \neq \emptyset$, necessarily $f_{i-1} = g_j$ on $D_{i-1} \cap \tilde{D}_j$. Hence, $g_j = f_{i-1} = f_i$ on $D_{i-1} \cap D_i \cap \tilde{D}_j$, but since $\tilde{D}_j \cap D_i$ is connected, we must have $g_j = f_i$ on the whole $\tilde{D}_j \cap D_i$, which is a contradiction. $\qquad\square$

Notice that an analytic function has a continuation from the center of a disk to the entire disk if and only if its power series at the center converges in the whole disk. Hence, the definition of analytic continuation along a curve also provides, at least in principle, a method to perform the continuation if it exists at all; cf. Exercise 19. In practice, however, one often can use some other representation of the continuation.

3.2 Example (The Gamma Function). The formula

$$\Gamma(z) = \int_0^\infty t^{z-1} e^{-t} dt$$

defines an analytic function for $\mathrm{Re}\, z > 0$ that is called the Gamma function. An integration by parts reveals that $z\Gamma(z) = \Gamma(z+1)$. Since $\Gamma(1) = 1$, it follows that $\Gamma(n) = (n-1)!$ for positive integers n and that $\Gamma(z)$ has a meromorphic continuation to the entire plane with poles at the points $0, -1, -2, \ldots$.

3.3 Example. Consider the function $f(z)$ defined in the unit disk by

$$f(z) = \frac{1}{\sqrt{\pi}} \sum_1^\infty \frac{1}{\sqrt{n}} z^n.$$

We claim that (f, U) can be continued along any curve from U that avoids the points 1 and 0 sic!. To begin with, as $\int_0^\infty \exp(-nt^2) dt = \sqrt{\pi/n}/2$, one finds that $f(z)$ coincides with

$$F(z) = \frac{2}{\pi} \int_0^\infty \frac{z}{e^{t^2} - z} dt \qquad (3.1)$$

in the unit disk, and therefore the formula (3.1) provides a continuation of f along any curve that does not intersect the half-axis $[1, \infty)$. Let $F_-(z)$ denote $F(z)$ below $[1, \infty)$ and $F_+(z)$ above. We claim that $F_-(z)$ can be continued across any point in $(1, \infty)$. To see this, use Cauchy's theorem and move the path of integration in (3.1) from, say, $1/R$ to R a little bit into the upper half-plane. If the modified path is sufficiently near the positive real axis, then its image under $\tau \mapsto \exp(\tau^2)$ will run in the upper half-plane, and hence the resulting integral will provide a continuation of F_- across the real axis between $\exp(1/R^2)$ and $\exp(R^2)$. In the same way, F_+ can be continued a little bit below $(1, \infty)$. By Exercise 23

$$\lim_{\epsilon \to 0} F(x + i\epsilon) - F(x - i\epsilon) = \frac{1}{\sqrt{\log x}} \qquad (3.2)$$

and hence, by uniqueness,

$$F_-(z) = F_+(z) - \frac{1}{\sqrt{\log z}}$$

in a neighborhood of $(1, \infty)$, where $\sqrt{\log z}$ is the branch that is positive on $(1, \infty)$. If it is continued from a point in $(1, \infty)$ along a closed curve that surrounds 1 but not 0, then one ends up with its negative. We leave it to the reader to determine the continuation of $f(z)$ along an arbitrary curve avoiding 0 and 1.

Analytic continuation plays an important role in Section 5 below.

3.4 The Monodromy Theorem. *Suppose that $\Omega \subset \mathbb{P}$ is simply connected and (f, D) is a function element that can be continued along any curve in Ω that starts at the center of D. Then there is a $g \in A(\Omega)$ such that $f = g$ in D.*

Proof. It is enough to show the following: If Γ and Γ_0 are curves that begin at the center of D and terminate at β, and g and g_0 are the continuations of f along Γ and Γ_0, respectively, then $g = g_0$. Assume the contrary. Since continuation along a curve is reversible, we thus have a closed curve $\gamma_0 = -\Gamma + \Gamma_0$ such that $g_0 \neq g$ is the continuation of g along γ_0. Since Ω is simply connected, there is a homotopy γ_t, $\gamma_t(0) = \gamma_t(1) = \beta$ (why?) between γ_0 and $\gamma_1 : [0, 1] \to \beta$. Let g_t be the continuation of g along γ_t. It is clear that $g_1 = g$. Now fix a t. Then there are $0 = s_0 < s_1 < \cdots < s_n = 1$ and (f_j, D_j) such that $\gamma_t([s_j, s_{j+1}]) \subset D_j$, $f_0 = g$, and $f_n = g_t$. Let ϵ be the least distance between $\gamma_t([s_j, s_{j+1}])$ and $\Omega \setminus D_j$, $j = 0, \ldots, n$. By compactness there is a $\delta > 0$ such that $|\gamma_t(s) - \gamma_u(s)| < \epsilon$ if $|u - t| < \delta$. Take such a u. Then (f_j, D_j) is also a continuation along γ_u and hence $g_u = g_t$. Thus $\{t \in [0, 1]; \ g_t = g\}$ is open and (trivially) closed and therefore equal to the whole interval $[0, 1]$. Hence $g_0 = g$, which contradicts our assumption. $\square$

3.5 Remark. What we in fact proved was: *If (f, D) can be continued along any curve in Ω that starts at the center of D, then the continuation along a closed null-homotopic curve must be f itself.* Incidentally, by considering, e.g., $\sqrt{\log z}$ in $\Omega = \mathbb{C} \setminus \{0, 1\}$, this proves that the curve in Example 3.8 in Ch. 1 is not null-homotopic.

§4. Simply Connected Domains

We now will sum up the various conditions on an open set that we have met thus far, which are equivalent to simple connectedness.

4.1 Theorem. *Suppose that $\Omega \subsetneq \mathbb{C}$ is connected. Then the following statements are equivalent:*

(a) *Ω is homeomorphic to the unit disk.*
(b) *Ω is simply connected.*
(c) *Any closed curve in Ω is null-homologous.*
(d) *$\mathbb{P} \setminus \Omega$ is connected.*
(e) *Ω is conformally equivalent to the unit disk.*
(f) *The monodromy theorem holds.*
(g) *Any $f \in A(\Omega)$ can be approximated u.c. in Ω by polynomials.*
(h) *$\int_\gamma f \, dz = 0$ for any closed γ and $f \in A(\Omega)$.*
(i) *Any $f \in A(\Omega)$ has a primitive function in Ω.*
(j) *Any nonvanishing $f \in A(\Omega)$ has a logarithm in Ω.*

In Ch. 4 we will add an additional equivalent condition, namely,
(k) *Each real harmonic function u in Ω is the real part of some $f \in A(\Omega)$.*

4.2 Remark. Notice that (a) to (d) are purely topological properties of an open set in $\mathbb{R}^2$, and they are still equivalent even if we remove the assumption that $\mathbb{C} \setminus \Omega \neq \emptyset$. It is clear that (a)$\rightarrow$ (b)$\rightarrow$ (c); cf. Proposition 3.7 in Ch. 1. However, it is not at all obvious that (c) implies (a). The corresponding statement for domains in higher dimensional $\mathbb{R}^N$ is not true. More precisely, there are connected domains with vanishing homology groups that are not contractible, as well as contractible domains that are not homeomorphic to the ball in $\mathbb{R}^N$.

Proof. Clearly, (a) $\rightarrow$ (b) $\rightarrow$ (c). From Cauchy's homology theorem it follows that (c) implies (h), and in Ch. 2 we saw that (h) $\rightarrow$ (i) $\rightarrow$ (j) $\rightarrow$ (e) $\rightarrow$ (a); therefore all of them are equivalent. Thus (d), (f), and (g) remain.

The Monodromy theorem states that (f) follows from (b). Conversely, suppose that (f) holds and take $w \in \mathbb{C} \setminus \Omega$. Any function element (f, D) in Ω, where $f(z)$ is some branch of $\log(z - w)$ in D, can be continued analytically along any curve in Ω. Thus if the monodromy theorem holds, there is a global function $g(z)$ that coincides with $f(z)$ on D. Hence, by uniqueness, $\exp g(z) = z - w$ in Ω so that $g'(z) = (z - w)^{-1}$; and hence

$$\mathrm{Ind}_\Gamma(w) = \frac{1}{2\pi i} \int_\Gamma \frac{dz}{z - w} = \frac{1}{2\pi i} \int_\Gamma g'(z) dz = 0$$

for all closed curves Γ. Thus all closed curves are null-homologous in Ω, and hence (f) is equivalent to the first group of conditions.

By Runge's theorem (more precisely, Corollary 1.6), (d) implies (g), which in turn implies (h). Hence, it just remains to see that (d) follows from the others. If (d) is false, then $\mathbb{P} \setminus \Omega$ consists of two disjoint compacts H and K, where K is compact in $\mathbb{C}$ and H contains ∞. Let $W = \Omega \cup K$. Since $W = \mathbb{P} \setminus H$, it is open and K is compact in W. Take $\phi \in C_0^\infty(W)$

with $\phi \equiv 1$ in a neighborhood of K, and choose $\alpha \in K$. Then

$$1 = -\frac{1}{\pi} \int \frac{\partial \phi}{\partial \bar{\zeta}} \frac{d\lambda(\zeta)}{\zeta - \alpha}. \tag{4.1}$$

Note that $\partial \phi / \partial \bar{\zeta}$ has its support in Ω. (Intuitively, one should think of $\partial \phi / \partial \bar{\zeta}$ as a curve in Ω with index 1 at α. Then (4.1) immediately contradicts (h).) If (i) holds, there is an analytic f in Ω such that $f' = (z - \alpha)^{-1}$, and hence

$$\int \frac{\partial \phi}{\partial \bar{\zeta}} f'(\zeta) d\lambda(\zeta) = -\int \frac{\partial^2 \phi}{\partial \bar{\zeta} \partial \zeta} f(\zeta) d\lambda(\zeta) = \int \frac{\partial \phi}{\partial \zeta} \frac{\partial f}{\partial \bar{\zeta}} d\lambda(\zeta) = 0,$$

which is a contradiction. $\qquad \square$

4.3 Remark. In the proof we used the fact that a compact set K in the plane is connected if and only if it has no nontrivial clopen (relatively closed and open) subset. It is *not* true that each component of K is clopen; for example, a Cantor set is totally disconnected, i.e., each component consists of just one single point. However, for each $a \in K$ the component of K containing a is the intersection of all clopen subsets of K that contain a.

§5. Analytic Functionals and the Fourier–Laplace Transform

A continuous linear functional μ on the space of entire functions $A(\mathbb{C})$, $\mu \in A'(\mathbb{C})$, is called an *analytic functional*. It is *carried* by the compact set K if for each open $\omega \supset K$ there is a constant C_ω such that

$$|\mu(f)| \leq C_\omega \sup_\omega |f|, \qquad f \in A(\mathbb{C}),$$

and any $\mu \in A'(\mathbb{C})$ is carried by some compact set K. In fact, since the family of sets

$$V_{K,\epsilon} = \{f \in A(\mathbb{C}); \sup_K |f| < \epsilon\}, \quad \epsilon > 0, \quad K \subset \mathbb{C}$$

is a basis for the topology on $A(\mathbb{C})$, $\mu^{-1}(U)$ must contain some $V_{K,\epsilon}$, and hence μ is carried by K. By the maximum principle it follows that if $\mathbb{P} \setminus K$ and $\mathbb{P} \setminus K'$ have the same unbounded component (the one containing ∞), then μ is carried by K if and only if it is carried by K'. Hence, it is natural to restrict to carriers with connected complements. Such compacts are called *polynomially convex* in $\mathbb{C}$ (cf. Exercise 5). For example, the functional

$$\mu(f) = \int_0^1 f(z) dz$$

is carried by any reasonable curve joining 0 and 1. For another example, let ϕ be analytic in $\mathbb{P} \setminus K$. Take a cut-off function χ that is identically 1 in a neigborhood of K, or take a set $\omega \supset K$ with reasonably smooth boundary, and let

$$\mu(f) = -\frac{1}{\pi} \int f\phi \frac{\partial \chi}{\partial \bar{z}} d\lambda(z) = \frac{1}{2\pi i} \int_{\partial \omega} f\phi dz. \tag{5.1}$$

Since this expression is independent of the particular choice of χ or ω (why), μ so defined is an analytic functional that is carried by K. Actually, any functional carried by K is of this kind. To see this, first notice that by the Hahn–Banach theorem μ can be extended to a functional on $C(\bar{\omega})$ and hence is represented by a measure $\tilde{\mu}$ on $\bar{\omega}$. Thus we can define

$$\phi(z) = \mu\left(\frac{1}{z - \cdot}\right) = \int \frac{d\tilde{\mu}(\zeta)}{z - \zeta},$$

and by varying ω this defines a ϕ that is analytic in $\mathbb{P} \setminus K$ and 0 at ∞. It is easily verified that ϕ represents μ in the sense (5.1).

For $\mu \in A'(\mathbb{C})$ we define the *Fourier–Laplace transform*

$$\hat{\mu}(z) = \mu\left(e^{z\cdot}\right).$$

This is an entire function and $\mu = 0$ if $\hat{\mu} = 0$. In fact, after differentiating and evaluating at 0 (keeping in mind that μ can be represented by some measure) it follows that μ vanishes on all polynomials, and the polynomials are dense in $A(\mathbb{C})$. If μ is carried by K, then

$$|\hat{\mu}(z)| \le C_\omega e^{H_\omega(z)}, \qquad \omega \supset K, \tag{5.2}$$

where $H_\omega(z)$ is the 1 homogeous function

$$H_\omega(z) = \sup_{\zeta \in \omega}(\operatorname{Re} z\zeta).$$

Recall that a compact set $K \subset \mathbb{C}$ is *convex* if its intersection with each line is connected. If K is convex, then

$$K = \{z; \ \operatorname{Re} z\zeta \le H_K(\zeta), \ \zeta \in \mathbb{C}\}. \tag{5.3}$$

This follows from the Hahn–Banach theorem since any real linear functional on $\mathbb{R}^2$ has the form $z \mapsto \operatorname{Re} \zeta z$; see also Exercise 36.

An entire function $h(z)$ that satisfies $|h(z)| \le C\exp(C|z|)$ is said to be of *exponential type*. Hence, the Fourier–Laplace transform of each $\mu \in A'(\mathbb{C})$ is of exponential type. However, the converse is also true. More precisely, we have

5.1 Theorem (Polya). *If $K \subset \mathbb{C}$ is compact and convex and $h(z)$ is an entire function satisfying (5.2), then $h(z)$ is the Fourier–Laplace transform of a unique $\mu \in A'(\mathbb{C})$ that is carried by K.*

Proof. The proof consists in constructing the desired analytic functional by means of the so-called Borel transform. Given an entire function f

of exponential type, the *Borel transform* is defined near ∞ (for z with $H_K(1/z) < 1$) by

$$Bf(z) = \frac{1}{z} \int_0^\infty f(\lambda/z)e^{-\lambda}d\lambda.$$

This clearly defines an analytic function in a neighborhood of ∞ that vanishes at ∞. We claim that if μ is defined via (5.1) ($\phi = Bf$), then $\hat{\mu} = f$. To this end first notice that $Bf \equiv 0$ implies that $f \equiv 0$ (differentiate $w \mapsto Bf(1/w)/w$ and evaluate at $w = 0$). Hence it is enough to show that if ϕ is analytic near ∞, say for $|z| > R/2$, and vanishes at ∞ and μ is the corresponding functional, then $B\hat{\mu} = \phi$. However, since

$$2\pi i \hat{\mu}(\zeta) = \int_{|w|=R} \phi(w)e^{\zeta w}dw,$$

we have for $|z| > R$ that

$$2\pi i B\hat{\mu}(z) = \frac{1}{z} \int_0^\infty \left(\int_{|w|=R} \phi(w)e^{\lambda w/z}dw \right) e^{-\lambda}d\lambda.$$

By the change of variables $w/z \mapsto w$ and Fubini's theorem the expression on the right-hand side equals

$$\int_{|w|=R/|z|} \phi(zw) \left(\int_0^\infty e^{\lambda(w-1)}d\lambda \right) dw = \int_{|w|=R/|z|} \frac{\phi(zw)dw}{1-w} = 2\pi i\phi(z),$$

where the last equality holds because ϕ is analytic in $|z| > R$ and vanishes at ∞.

To see that μ is carried by K if K is convex, we have to verify that Bf has an analytic continuation to the set $\mathbb{P} \setminus K$. For any real θ let Γ_θ be the half-line $0 \le t \mapsto e^{i\theta}t$ and let

$$B_\theta f(z) = \int_{\Gamma_\theta} f(\lambda)e^{-\lambda z}d\lambda = \int_0^\infty f(e^{i\theta}t)e^{-e^{i\theta}tz}e^{i\theta}dt.$$

Since $\mathrm{Re}(e^{i\theta}z) > H_K(e^{i\theta})$ if and only if the same equality holds for some $\omega \supset K$ instead of K, the condition (5.2) implies that $B_\theta f$ is analytic in the half-plane $\Pi_\theta = \{\mathrm{Re}\, e^{i\theta}z > H_K(e^{i\theta})\}$. For all large z with $e^{i\theta}z$ real and positive, a simple change of variable reveals that $B_\theta f(z) = Bf(z)$. Hence they coincide on their common set of definition, and hence $Bf(z)$ has a continuation to the union of all the sets Π_θ which, in view of (5.3), is precisely $\mathbb{P} \setminus K$. $\qquad\square$

We conclude with a related result.

5.2 Theorem (Paley–Wiener). *Suppose that $f(z)$ is an entire function such that*

$$|f(z)| \le Ce^{A|z|}$$

and

$$\int_{-\infty}^{\infty} |f(x)|^2 dx < \infty. \tag{5.4}$$

Then there is a $g(t) \in L^2(-A, A)$ such that

$$f(z) = \int_{-A}^{A} g(\xi) e^{iz\xi} d\xi. \tag{5.5}$$

Note that the converse is immediate in view of Plancherel's theorem since $f(-x)$ is the Fourier transform of g.

Proof. As in the preceding proof, we find that the Borel transform $Bf(z)$ is analytic in $\{|z| > A\}$. However, in view of the assumption (5.4), it follows that in fact $B_0 f(z)$ is defined for $\operatorname{Re} z > 0$ and $B_\pi f$ for $\operatorname{Re} z < 0$, and therefore $Bf(z)$ is analytic in the complement of the interval $[-iA, iA]$. Now let $f_\epsilon(x) = f(x) e^{-\epsilon|x|}$. Then

$$\hat{f}_\epsilon(\xi) = \int_{-\infty}^{\infty} f_\epsilon(t) e^{-it\xi} dt = B_0 f(\epsilon + i\xi) - B_\pi f(-\epsilon + i\xi),$$

which tends to 0 when $\epsilon \to 0$ if $|\xi| > A$. On the other hand, since $f_\epsilon \to f$ in L^2, Plancherel's theorem implies that $\hat{f}_\epsilon \to \hat{f}$ in L^2, and hence $\hat{f}(\xi) = 0$ for $|\xi| > A$. If $g = \hat{f}$, then (5.5) holds for real z by the inversion formula, and as both sides are entire functions, it holds for all z. $\qquad \square$

§6. Mergelyan's Theorem

Let K be a compact set in the plane and suppose that f is a complex function on K that can be uniformly approximated by analytic polynomials on K. It then follows that f is continuous on K and analytic in the interior. If any such f can be approximated uniformly by polynomials, then the complement of K must be connected; cf. the discussion after Runge's theorem.

6.1 Mergelyan's Theorem. *Let K be a compact set in the plane such that the complement is connected, and suppose that f is continuous on K and analytic in the interior of K. To each $\epsilon > 0$ there is a polynomial such that $|f - p| < \epsilon$ on K.*

Notice that Runge's theorem applies only if f is analytic in a neighborhood of K, and therefore Mergelyan's theorem is considerably stronger. In particular, if the interior of K is empty, any continuous function can be approximated uniformly by analytic polynomials. If K is an interval, this is the classical Weierstrass' theorem. This section is devoted to the proof of Mergelyan's theorem.

Proof. To begin with, we can extend f to a continuous function with compact support in $\mathbb{C}$, which we also denote f. This is an application of Tietze's extension theorem. Let $\omega(\delta)$ be the modulus of continuity of f,

$$\omega(\delta) = \sup\{|f(z) - f(w)|; \ |z - w| < \delta\}.$$

Since f is uniformly continuous, $\omega(\delta) \to 0$ when $\delta \to 0$. Hence, it is enough to find, for each δ, a polynomial p such that

$$|f(z) - p(z)| \le C\omega(\delta), \quad z \in K, \tag{6.1}$$

where C is independent of δ. In what follows, C denotes such a constant, but it can be different in different places.

Let ϕ be a smooth positive function with support in $D(0, 1/2)$ such that $\int \phi d\lambda = 1$, and let $\phi_\delta(z) = \delta^{-2}\phi(z/\delta)$. The function

$$\Phi(z) = f * \phi_\delta(z) = \int \phi_\delta(z - w)f(w)d\lambda(w) = \int \phi_\delta(w)f(z - w)d\lambda(w)$$

is then smooth, and since

$$f(z) - \Phi(z) = \int (f(z) - f(z - w))\phi_\delta(w)d\lambda(w),$$

it follows that

$$|f(z) - \Phi| \le \omega(\delta). \tag{6.2}$$

Moreover,

$$\frac{\partial \Phi}{\partial \bar{z}} = \int \frac{\partial \phi_\delta}{\partial \bar{w}}(w)f(z - w)d\lambda(w) = \int \frac{\partial \phi_\delta}{\partial \bar{w}}(w)\big(f(z - w) - f(z)\big)d\lambda(w)$$

(since $\int (\partial \phi_\delta / \partial \bar{w})(w)d\lambda(w) = 0$), and hence we get the estimate

$$\left|\frac{\partial \Phi}{\partial \bar{z}}\right| \le \frac{C\omega(\delta)}{\delta}, \tag{6.3}$$

since $\int |\partial \phi_\delta / \partial \bar{w}| d\lambda(w) \le C/\delta$. Thus, we have approximated f so far by the function Φ, which at least is analytic at points in K that have distance more than δ to ∂K. Let H denote the support of $\partial \Phi / \partial \bar{w}$. The crucial part of the proof is contained in the following proposition.

6.2 Proposition. *There is an open neighborhood Ω of K and a continuous function $r(\zeta, z)$ defined for $\zeta \in H$ and $z \in \Omega$ such that $r(\zeta, z)$ is holomorphic for $z \in \Omega$,*

$$\left|r(\zeta, z) - \frac{1}{\zeta - z}\right| \le \frac{C\delta^2}{|\zeta - z|^3},$$

and

$$|r(\zeta, z)| \le \frac{C}{\delta},$$

where the constant C is independent of δ.

Taking this proposition for granted, it is now easy to conclude the proof of the theorem. By formula (1.7) in Ch. 1 we have that

$$\Phi(z) = -\frac{1}{\pi} \int_H \frac{(\partial\Phi/\partial\bar{\zeta})d\lambda(\zeta)}{\zeta - z}.$$

Now the function

$$F(z) = -\frac{1}{\pi} \int_H r(\zeta, z)\frac{\partial\Phi}{\partial\bar{\zeta}}d\lambda(\zeta)$$

is analytic in $\Omega \supset K$, and by (6.3)

$$|F(z) - \Phi(z)| \le \frac{C\omega(\delta)}{\delta} \int_H \left| r(\zeta, z) - \frac{1}{\zeta - z} \right| d\lambda(\zeta).$$

In this integral we estimate the integrand by $C/\delta + |\zeta - z|^{-1}$ when $|\zeta - z| \le \delta$ and by $C\delta^2/|\zeta - z|^3$ when $|\zeta - z| > \delta$. In any case, we get the estimate $C\omega(\delta)$ where C is independent of δ. In combination with (6.2) we get that $|f(z) - F(z)| \le C\omega(\delta)$ on K, and since F is analytic in a neighborhood of K, we can apply Runge's theorem and obtain a polynomial p such that (6.1) holds. □

It remains to prove Proposition 6.2, and to this end we need yet another result.

6.3 Proposition. Let D be a disk with radius δ and E a connected compact subset with diameter at least δ such that $\mathbb{P} \backslash E$ is also connected. Then there is a smooth function $r(\zeta, z)$ defined for $z \in \mathbb{P} \backslash E$ and $\zeta \in D$ that is analytic in z and satisfies

$$\left| r(\zeta, z) - \frac{1}{\zeta - z} \right| \le \frac{C\delta^2}{|\zeta - z|^3} \tag{6.4}$$

and

$$|r(\zeta, z)| \le \frac{C}{\delta}, \tag{6.5}$$

where the constant C is independent of δ.

Proof. By a change of scale and a translation we may assume that $\delta = 1$ and that D is the unit disk U. Note that in particular it is necessary to find an analytic function $g(z)$ in $\mathbb{P} \backslash E$ that is bounded, and such that

$$zg(z) \to 1 \quad \text{when} \quad z \to \infty. \tag{6.6}$$

The latter condition means that $g(\infty) = 0$ and that its derivative at ∞ is 1 with respect to the coordinate $w = 1/z$. It follows from the Riemann mapping theorem and its proof that such functions exist, and that if $g(z)$ is such a function with minimal sup norm, then $g(z)$ is in fact a bijection onto some disk $D(0, t)$. Now Proposition 3.5 in Ch. 2 applies to the inverse of the function $z \mapsto g(z/t)/t$, and hence the diameter of the set tE is less than 4; so $t < 4$ and thus $|g(z)| < 4$.

For fixed $\zeta \in U$ and $|z - \zeta| > 2$ we have

$$g(z) = \frac{1}{z - \zeta} + \frac{a_2(\zeta)}{(z - \zeta)^2} + \mathcal{O}\left(\frac{1}{|z - \zeta|^3}\right).$$

If we define

$$r(\zeta, z) = g(z) - a_2(\zeta)g^2(z), \tag{6.7}$$

then

$$r(\zeta, z) - \frac{1}{z - \zeta} = \mathcal{O}\left(\frac{1}{|z - \zeta|^3}\right)$$

when $z \to \infty$. Now

$$a_2(\zeta) = \frac{1}{2\pi i} \int_{|z| = R} (z - \zeta)g(z)dz = b - \zeta,$$

where $b = (1/2\pi i) \int_{|z| = R} zg(z)dz$. We can change the path of integration to the unit circle, and since $|g| \leq 4$, we then get that $|b| \leq 4$ and hence also the estimate (6.5). Finally, the function

$$z \mapsto (z - \zeta)^3 \left(r(\zeta, z) - \frac{1}{z - \zeta}\right)$$

is analytic in $\mathbb{P} \setminus E$ (it is bounded when $z \to \infty$ and hence it has removable singularity at ∞), and it is bounded by some constant C in $U \setminus E$ that is independent of $\zeta \in U$; therefore, by the maximum principle it is bounded by the same constant in $\mathbb{P} \setminus E$. Thus (6.4) holds. $\square$

Proof of Proposition 6.2. We can cover H by a finite number of disks D_j with radii 2δ and centers outside K. Moreover, since the complement of K is connected, in each D_j we can find a set E_j of diameter at least 2δ that does not intersect K (there must be a curve from the center to the boundary that does not intersect K). For each D_j let $r_j(\zeta, z)$ be the functions given by Proposition 6.3, and let $\Omega = \cap \mathbb{P} \setminus E_j$. Then clearly $\Omega \supset K$, and if ϕ_j is a partition of unity subordinate to the open cover D_j of the compact set H, then

$$r(\zeta, z) = \sum_j \phi_j(\zeta)r_j(\zeta, z)$$

has the required properties. $\square$

6.4 Remark. Essentially the same proof also gives the following more general result: *Suppose that $K \subset \mathbb{P}$ is compact and $\mathbb{P} \setminus K$ has a finite number of components. Choose one point a_j from each component. Then any $f \in C(K) \cap A(\mathrm{int}(K))$ can be uniformly approximated by rationals with poles only at the points a_j.*

Supplementary Exercises

Exercise 1. Show that if $\Omega \subset \mathbb{C}$ is bounded, then $\mathbb{P} \setminus \Omega$ is connected if and only if $\mathbb{C} \setminus \Omega$ is. Give a counterexample when Ω is unbounded.

Exercise 2. Let $\Omega = \{z; \, |z| < 1, \, |2z - 1| > 1\}$ and $f \in A(\Omega)$. Show that there are polynomials p_n such that $p_n \to f$ u.c. in Ω. What can be said about f if $p_n \to f$ uniformly in Ω?

Exercise 3. Show that there are polynomials p_n such that $\lim p_n = 0$ in $\mathbb{C} \setminus \{0\}$ and $p_n(0) = 1$.

Exercise 4. Do polynomials p_n exist such that $\lim p_n(z)$ is equal to 1 when $\mathrm{Im}\, z > 0$, 0 when $\mathrm{Im}\, z = 0$, and -1 when $\mathrm{Im}\, z < 0$?

Exercise 5. For a compact $K \subset \mathbb{C}$, define the polynomially convex hull $\widehat{K} = \{z \in \mathbb{C}; \, |p(z)| \leq \sup_K |p| \text{ for all polynomials } p\}$. Note that $\widehat{K} \supset K$. A compact set K is said to be *polynomially convex* if $\widehat{K} = K$. Show that K is polynomially convex if and only if $\mathbb{C} \setminus K$ is connected.

Exercise 6. Supply the details in the following constructive proof of Runge's theorem. First consider the set $A = \{\alpha \in \mathbb{C} \setminus K$; the functions $z \mapsto (z - \alpha)^{-k}$, $k = 1, 2 \ldots$, can be uniformly approximated on K by rationals with poles in $\{a_j\}\}$. A is trivially relatively closed in $\mathbb{C} \setminus K$. Take $\alpha \in A$. For β near α

$$\frac{1}{z - b} = \sum_0^\infty \frac{(\alpha - \beta)^k}{(z - \alpha)^{k+1}}$$

with uniform convergence on K. Hence $(z - \beta)^{-1}$ can be approximated on K by our rationals and hence also $(z - \beta)^{-k}$ for $k \geq 1$. Thus A is open. Since A intersects each component of $\mathbb{C} \setminus K$, it follows that $A = \mathbb{C} \setminus K$. If f is analytic in a neighborhood of K, as usual one can express f as a Cauchy integral where the integration is performed outside K, and this integral then can be approximated uniformly on K by a finite Riemann sum in which the terms are functions of the type $z \mapsto (z - \alpha)^{-1}$ for $\alpha \in \mathbb{C} \setminus K$.

Exercise 7. Use Theorem 2.3 to prove the statement in Remark 2.2. Hint: Let ϕ_j be a locally finite partition of unity subordinate to the open cover Ω_k (i.e., locally all but a finite number of the ϕ_j vanish, $\sum \phi_j \equiv 1$, and to each j there is an n_j such that $\phi_j \in C_0^\infty(\Omega_{n_j})$). Let $g_k = \sum(f_k - f_{n_j})\phi_j$. Then

g_k is smooth in Ω_k and $g_k - g_\ell = f_k - f_\ell$ in $\Omega_k \cap \Omega_\ell$. Now $\psi = \partial g_k/\partial \bar{z}$ defines a global smooth function in Ω, and if $\partial u/\partial \bar{z} = \psi$ in Ω, then $f = f_k - g_k + u$ is a global meromorphic function with the required properties.

Exercise 8. Show that for each $\delta > 0$ and $\gamma > 0$ there is a positive constant $C_{\gamma,\delta}$ such that

$$\int_U \frac{(1 - |\zeta|)^{-1+\delta}}{|1 - \bar{\zeta}z|^{1+\delta+\gamma}} \leq C_{\gamma,\delta} \frac{1}{(1 - |z|)^\gamma}. \tag{*}$$

Exercise 9. Here is an outline of a proof of the statements in Example 2.5.
(i) Prove that P is bounded on $L^2(U)$. By Schwarz' inequality,

$$|Pu(z)|^2 \leq C \int_U \frac{(1 - |\zeta|^2)^{-1/2}}{|1 - \bar{\zeta}z|^2} \int_U \frac{(1 - |\zeta|^2)^{1/2}|u(\zeta)|^2}{|1 - \bar{\zeta}z|^2}.$$

Then apply (*) to the first integral on the right-hand side and use Fubini's theorem.
(ii) Show that the space of analytic polynomials is dense in $B(U)$. Show first that if $f \in B(U)$ is orthogonal to all polynomials, then $f = 0$.
(iii) Verify (2.3).

Exercise 10. Are the polynomials dense in $B^p = A(U) \cap L^p(U)$?

Exercise 11. Suppose that Ω is bounded, $\psi(\zeta, z) \in C^1(\Omega \times \Omega)$, $z \mapsto \psi(\zeta, z)$ is analytic for each fixed $\zeta \in \Omega$, and $\psi(\zeta, \zeta) = 1$. Show that

$$u(z) = -\frac{1}{\pi} \int_\Omega \psi(\zeta, z) \frac{f(\zeta) d\lambda(\zeta)}{\zeta - z} \tag{**}$$

is a solution to $\partial u/\partial \bar{z} = f$ if $f \in C^1(\Omega)$ and for each $K \subset \Omega$ there is a constant C_K such that

$$\sup_{z \in K} \int_\Omega |\psi(\zeta, z) f(\zeta)| d\lambda(\zeta) \leq C_K.$$

Exercise 12. Also show the converse: If $\psi \in C^1(\Omega \times \Omega)$ and (**) is a solution for all f in, for example, $C_0^\infty(\Omega)$, then $z \mapsto \psi(\zeta, z)$ is analytic and $\psi(\zeta, \zeta) = 1$.

Exercise 13. Show that

$$K_\alpha f(z) = -\frac{1}{\pi} \int_{|\zeta|<1} \left(\frac{1 - |\zeta|^2}{1 - \bar{\zeta}z} \right)^\alpha \frac{f(\zeta) d\lambda(\zeta)}{\zeta - z}$$

solves $\partial u/\partial \bar{z} = f$ in the unit disk if $|f(\zeta)| \leq C(1 - |\zeta|^2)^{-\alpha+\epsilon-1}$.

Exercise 14. Use the operators K_α to show that there is a solution to (2.2) in U with $|u(z)| \leq C(1 - |z|)^{-r}$ if $|f(\zeta)| \leq (1 - |\zeta|)^{-r-1}$, $r > 0$.

Exercise 15. For $\alpha > 0$, let L_α^2 be L^2 in U with respect to the weight $(1 - |\zeta|^2)^{\alpha-1}d\lambda(\zeta)$. Let $B_\alpha = A(U) \cap L_\alpha^2$ and

$$P_\alpha u(z) = \frac{\alpha}{\pi} \int_U \frac{(1 - |\zeta|^2)^{\alpha-1} u(\zeta) d\lambda(\zeta)}{(1 - \bar\zeta z)^{\alpha+1}}.$$

Try to generalize Example 2.5.

Exercise 16. Let $\Omega = \mathbb{C}$ and take ψ in Exercise 11 as an integer power of $(1 + \bar\zeta z)/(1 + |\zeta|^2)$. Are the occurring solutions minimal in some sense? What are the corresponding projection operators? Consider the same questions for $\psi = \exp(\bar\zeta z)$.

Exercise 17. One can obtain Mittag–Leffler's theorem directly from Theorem 2.3: First construct a $\psi \in C^\infty(\Omega \setminus \{a_j\})$ such that $\psi - p_j$ is analytic in a neighborhood of a_j for each j. Show that $g = (\partial/\partial\bar z)(\psi - p_j)$ is a global C^∞ function in Ω. Take $u \in C^\infty(\Omega)$ such that $\partial u/\partial\bar z = g$ and consider $f = \psi - u$.

Exercise 18. Let $f \in C_0^\infty(\mathbb{C})$ have support K (or, more generally, take a measure μ on K). Let a_j be a set of points, one from each component of $\mathbb{P} \setminus K$. Show that $\partial u/\partial\bar z = f$ has a solution with support in K if and only if $\int_K r(z) f(z) d\lambda(z) = 0$ for all rationals with poles in the set $\{a_j\}$. Give an interpretation of the proof of Runge's theorem in the sense of distributions.

Exercise 19. Suppose that $a_n \geq 0$ and $\limsup(a_n)^{1/n} = 1$. Show that 1 is a singular point for $f(z) = \sum a_n z^n$, i.e., show that $f(z)$ cannot be continued along the positive real axis past the point 1. Hint: Consider the power series expansion of f around the point $1/2$.

Exercise 20. Assume that $(\tilde f, \tilde D)$ is the analytic continuation of (f, D) along some curve γ. Let g be an entire function, and suppose that $g \circ f \equiv 0$ in D. Show that $g \circ \tilde f \equiv 0$ in $\tilde D$.

Exercise 21. Compute $\Gamma(m/2)$ for integers m. What are the residues at the points $0, -1, -2, \ldots$?

Exercise 22. Let $\phi \in C^\infty(\bar U)$, and define

$$f(\alpha) = \frac{\alpha}{\pi} \int_{|\zeta|<1} (1 - |\zeta|^2)^{\alpha-1} \phi(\zeta) d\lambda(\zeta).$$

Show that $f(\alpha)$ is analytic for $\operatorname{Re} \alpha > 0$ and show that f can be continued to a meromorphic function in the whole plane with (possible) poles only at $\alpha = -1, -2, \ldots$. In particular, it has a removable singularity at $\alpha = 0$. Show that if ϕ is holomorphic, then $f(\alpha)$ is constantly equal to $\phi(0)$.

Exercise 23. Verify the limit (3.2). Hint: Let $f(t)$ be a locally integrable function on $\mathbb{R}$ such that $\int_{-\infty}^\infty |f(t)|dt/(1 + t^2) < \infty$ and define the Poisson integral

$$Pf(y, x) = \frac{1}{\pi} \int_{-\infty}^\infty \frac{yf(t)dt}{y^2 + (x - t)^2}, \qquad y > 0.$$

Then $Pf(y, x) \to f(x)$ when $y \to 0$ if $f(t)$ is continuous at $t = x$.

Exercise 24. Let K be a connected compact set in $\mathbb{C}$. Show that $\mathbb{C} \setminus K$ is connected if and only if K has a neighborhood basis consisting of simply connected (open) sets. A family of open neighborhoods of K is a neighborhood basis if any neighborhood of K contains some set from the family.

Exercise 25. Let D be a connected domain in $\mathbb{C}$. Show that the equation $p(\partial)u = f$ has an analytic solution in D for each $f \in A(D)$ and each constant coefficient linear differential operator

$$p(\partial) = \sum_{k=1}^{m} a_k \frac{\partial^k}{\partial \zeta_k},$$

if and only if D is simply connected.

Exercise 26. Suppose that $f \in A(\Omega)$, Ω is connected, $f \not\equiv 0$, and that for each positive integer n there is a $g \in A(\Omega)$ such that $g^n = f$. Show that f has a logarithm in Ω. Hint: $\log f$ exists if and only if $\int_\gamma f'dz/f = 0$ for all closed curves γ in Ω.

Exercise 27. Find all $\mu \in A'(\mathbb{C})$ that are carried by the set $\{0\}$.

Exercise 28. Show that if $\mathbb{P} \setminus K$ is connected, then there is a 1-1 correspondence between the space of functionals carried by K and the set of analytic functions in $\mathbb{P} \setminus K$ that vanish at ∞.

Exercise 29. Suppose that $\Omega \supset K$ and that $\phi \in A(\Omega \setminus K)$. Let $\{a_j\}$ contain one point from each component of $\mathbb{P} \setminus K$, and let $K \subset \omega \subset\subset \Omega$ such that ω is disjoint from $\{a_j\}$. Show that ϕ is the restriction of a function $\Phi \in A(\Omega)$ if and only if $\int_{\partial \omega} r\phi dz = 0$ for all rationals r with poles in $\{a_j\}$. Hint: Take a function $F \in C^\infty(\Omega)$ that coincides with ϕ in the complement of a neighborhood of K. Modify F to an analytic function by solving $\partial u/\partial \bar{z} = \partial F/\partial \bar{z}$ in an appropriate way.

Exercise 30. Let $f(z) = \sum_0^\infty a_n z^n$ be of exponential type. Show that the Borel transform is given by

$$Bf(z) = \sum_0^\infty a_n n! z^{-(n+1)}$$

for large $|z|$. Use this to prove directly that $\hat{\mu} = f$ if $\phi = Bf$ and (5.1) holds.

Exercise 31. Suppose that $\mu \in A'(\mathbb{C})$ is carried by K. Show that in each open $\omega \supset K$ there is a sequence of points $a_j \in \omega$ and numbers c_j such that $\mu = \sum c_j \delta_{a_j}$.

Exercise 32. It is clear that $\{z; H_K(1/z) < 1\} \subset \mathbb{P} \setminus K$. For which convex K is the inclusion strict?

Exercise 33. Use Cauchy's theorem to show that all of the $B_\theta(z)$ in Section 5 coincide on their overlaps.

Exercise 34. Verify the statement in Remark 6.4.

Exercise 35. Give an elementary proof of Mergelyan's theorem when K is the closed unit disk (one cannot use the Taylor polynomials directly).

Exercise 36. Assume that $K \subset \mathbb{C}$ is convex. Show that through each point $p \in \mathbb{C} \setminus K$ there is some line that does not intersect K. Use this result to conclude (5.3). Hint: Assume that $0 \notin K$ and that each line through 0 intersects K. For each real θ, the line $t \mapsto e^{i\theta}t$ intersects K either for some positive or some negative t but not both. Show that the set of θ for which the intersection occurs for positive t is both open and closed. Deduce a contradiction.

Notes

Runge's theorem was published in 1885. The Poisson integral in Exercise 23 solves Dirichlet's problem in the upper half-plane; in Ch. 4, the corresponding integral in the unit disk is studied.

Polya's theorem (Theorem 5.1) is from 1929; see [B] for references and further results about functions of exponential type.

Mergelyan's theorem is from 1952; see AMS Translations 101 (1954). For a different proof see Carleson, *Math. Scand.*, vol. 15 (1964).

4

Harmonic and Subharmonic Functions

§1. Harmonic Functions

A function $u \in C^2(\Omega)$ is *harmonic* in Ω if $\Delta u = 0$. If u, v are harmonic and $\alpha, \beta \in \mathbb{C}$, then $\alpha u + \beta v$ is harmonic. If $f \in A(\Omega)$, then f, $\bar{f}$, $\operatorname{Re} f$, and $\operatorname{Im} f$ are harmonic in Ω. If, in addition, f is nonvanishing, then $\log |f|$ is harmonic. If $\Omega = \mathbb{C} \setminus \{0\}$, then $\log |z|$ is harmonic in Ω, but there is no $f \in A(\Omega)$ such that $\log |z| = \operatorname{Re} f$. However, we have

1.1 Proposition. *If u is real and harmonic in a simply connected domain Ω, then $u = \operatorname{Re} f$ for some $f \in A(\Omega)$.*

Proof. Since $2\partial u/\partial z \in A(\Omega)$ and Ω is simply connected, there is a $g \in A(\Omega)$ such that $g' = \partial g/\partial z = 2\partial u/\partial z$. Thus $\partial(g + \bar{g})/\partial z = 2\partial u/\partial z$, and since u is real, $2u = g + \bar{g} + 2c$ for some real constant c; therefore, $u = \operatorname{Re}(g + c)$. □

In particular, we find that harmonic functions are C^∞. Moreover, we have the following uniqueness property: *If u is harmonic in a connected set Ω and vanishes in an open subset, then u vanishes identically in Ω.* In fact, one may assume that u is real. Then the statement follows from the identity theorem for analytic functions applied to $\partial u/\partial z$. If u is harmonic in a neighborhood of $\overline{D(z, r)}$, then

$$u(z) = \frac{1}{2\pi} \int_0^{2\pi} u(z + re^{i\theta})d\theta, \tag{1.1}$$

i.e., u has the *mean value property*. Again we may assume that u is real, and then (1.1) follows from Proposition 1.1 and the mean value property for analytic functions. From (1.1) and the uniqueness property it follows that a real harmonic u has no local maxima or minima unless it is locally constant. Analogously to Corollary 1.13 in Ch. 1, we also have the following.

1.2 The Maximum Principle. *If Ω is bounded and $u \in C(\overline{\Omega})$ is real and harmonic in Ω, then u attains its maximum and minimum on $\partial\Omega$. In particular, $u \equiv 0$ if $u = 0$ on $\partial\Omega$.*

We now shall derive a generalization of (1.1) that represents the value at an arbitrary interior point in terms of the boundary values on the disk. If u is real and harmonic in a neighborhood of $\overline{U}$, then, by Proposition 1.1,

$$u(z) = \frac{1}{2}\left(\sum_0^\infty a_n z^n + \sum_0^\infty \bar{a}_n \bar{z}^n\right) = \sum_{-\infty}^\infty A_n r^{|n|} e^{in\theta} \qquad (1.2)$$

(where $z = re^{i\theta}$) with absolute convergence on $\overline{U}$. Hence,

$$u(re^{i\theta}) = \frac{1}{2\pi}\int_0^{2\pi} u(e^{it})\sum_{-\infty}^\infty r^{|n|}e^{in(\theta-t)}dt, \quad r < 1$$

(replace $re^{i\theta}$ with e^{it} in (1.2), plug it into the integral above, and use the fact that $\int e^{imt}dt = 0$ for nonzero integers m), so that

$$u(re^{i\theta}) = \frac{1}{2\pi}\int_0^{2\pi} u(e^{it})P_r(\theta-t)dt = \frac{1}{2\pi}\int_0^{2\pi} u(e^{i(\theta-t)})P_r(t)dt, \ 0 \le r < 1,$$

where $P_r(t) = \sum_{-\infty}^\infty r^{|n|}e^{int}$. If $z = re^{i\theta}$ and $\zeta = e^{it}$, then the *Poisson kernel* is

$$P_r(\theta - t) = \sum_{-\infty}^\infty r^{|n|}e^{in(\theta-t)} = \text{Re}\left(\frac{e^{it} + z}{e^{it} - z}\right)$$

$$= \frac{1 - r^2}{1 - 2r\cos(\theta-t) + r^2} = \frac{1 - |z|^2}{|1 - \bar{\zeta}z|^2} = \frac{1 - |z|^2}{|\zeta - z|^2}.$$

Verify the equalities! The second one reveals that the Poisson kernel is a harmonic function of $z = re^{i\theta}$ for fixed e^{it}.

Exercise 1. Show that $P_r(t) = P_r(-t)$, $t \mapsto P_r(t)$ is strictly decreasing on $(0, \pi)$, $P_r(\delta) \to 0$ when $r \nearrow 1$ for fixed $\delta > 0$, and that $(1/2\pi)\int P_r(t) = 1$ for all $r < 1$; cf. Figure 1 in Section 1 of Ch. 6.

1.3 Remark. Since harmonicity is preserved under translation and dilation, we get that

$$u(a + re^{i\theta}) = \frac{1}{2\pi}\int_0^{2\pi} \frac{R^2 - r^2}{R^2 - 2Rr\cos(\theta - t) + r^2}u(a + Re^{it})dt, \quad r < R,$$

if u is harmonic in a neighborhood of $\overline{D(a, R)}$.

For $f \in C(T)$, one defines the *Poisson integral* Pf of f as

$$Pf(re^{i\theta}) = \frac{1}{2\pi} \int_0^{2\pi} f(e^{it}) P_r(\theta - t) dt$$

$$= \frac{1}{2\pi} \int_0^{2\pi} f(e^{i(\theta-t)}) P_r(t) dt, \quad r < 1,$$

which is a harmonic function in U since the Poisson kernel is harmonic.

1.4 Proposition. *If $f \in C(T)$, then Pf is harmonic in U; and if Pf is defined as f on T, then $Pf \in C(\overline{U})$.*

Dirichlet's problem is the following: Given a function on $\partial\Omega$, find a harmonic function F in Ω such that $F = f$ on $\partial\Omega$. To discuss its solvability, one of course must specify which class of domains and functions f one allows, and the exact meaning of $f = F$. For bounded domains Ω with reasonable boundary (e.g., piecewise C^1), any continuous f on $\partial\Omega$ has a continuous harmonic extension F to Ω, which in view of the maximum principle is necessarily unique. In this book we essentially will restrict our attention to the unit disk. Proposition 1.4 says that the solution to Dirichlet's problem in U for continuous f is given by the Poisson integral Pf.

Proof of Proposition 1.4. We shall prove that $Pf(re^{i\theta}) \to f(e^{i\theta})$ uniformly in θ when $r \nearrow 1$. Given $\epsilon > 0$, take $\delta > 0$ such that $|f(e^{i(\theta-t)}) - f(e^{i\theta})| < \epsilon$ if $|t| < \delta$. Now

$$Pf(re^{i\theta}) - f(e^{i\theta}) = \frac{1}{2\pi} \int_0^{2\pi} \left(f(e^{i(\theta-t)}) - f(e^{i\theta}) \right) P_r(t) dt.$$

In the set $\delta \leq |t| \leq \pi$, $P_r(t) \leq P_r(\delta)$ (cf. Exercise 1), whereas in the set $0 \leq |t| < \delta$, $|f(e^{i(\theta-t)}) - f(e^{i\theta})| < \epsilon$. Since $P_r(t)/2\pi$ is positive and has integral 1, we get the estimate

$$|Pf(re^{i\theta}) - f(e^{i\theta})| \leq 2\|f\|_\infty P_r(\delta) + \epsilon,$$

which is less than 2ϵ if r is sufficiently close to 1.   □

We now consider a converse of the mean value property.

1.5 Theorem. *Suppose that $u \in C(\Omega)$ and that, for each $z \in \Omega$, (1.1) holds for small $r > 0$. Then u is harmonic in Ω.*

Proof. We may assume that u is real. Moreover, it is enough to prove that u is harmonic in each disk $D \subset\subset \Omega$. Given such a D, take the harmonic function h such that $h = u$ on ∂D. Then $v = u - h$ also satisfies the hypothesis in D. If v had a positive supremum in D, then it would be attained on a nonempty compact set $K \subset D$, and this in turn would violate (1.1) for $z \in K$ with minimal distance to ∂D. Thus, $v \leq 0$ in D. For the same reason, $v \geq 0$, and therefore $u = h$ in D. □

The proof above shows that it is enough to assume that for each $z \in \Omega$ there is a sequence $r_n \searrow 0$ for which (1.1) holds. From Theorem 1.5 it follows that u is harmonic if $u_n \to u$ u.c. and each u_n is harmonic.

1.6 Weyl's Lemma. *If $u \in L^1_{\text{loc}}(\Omega)$ (or $u \in \mathcal{D}'(\Omega)$) and $\int u \Delta \phi = 0$ for all $\phi \in C_0^\infty$, then u is harmonic; more precisely, there is a harmonic function v in Ω such that $u = v$ a.e. (or as distributions).*

We will not rely on this result and leave the proof as an exercise; it is in fact a consequence of Theorem 2.12 below.

1.7 Harnack's Theorem. *If u_n are harmonic in a connected domain Ω and $u_1 \leq u_2 \leq u_3 \ldots$, then either $u_n \nearrow \infty$ for all $z \in \Omega$ or $u_n \nearrow u < \infty$ u.c. in Ω.*

Proof. Take $\overline{D(a, R)} \subset \Omega$. We may assume that $u_1 \geq 0$. Since

$$(R - r)^2 \leq R^2 - 2rR\cos(\theta - t) + r^2 \leq (R + r)^2,$$

we have, by Remark 1.3,

$$\frac{R - r}{R + r} u_n(a) \leq u_n(a + re^{i\theta}) \leq \frac{R + r}{R - r} u_n(a), \quad r < R.$$

Thus, if $u_n(a) \nearrow \infty$, then $u_n \nearrow \infty$ in a neighborhood of a. On the other hand, if $u_n(a) \nearrow u(a) < \infty$, then $u = \lim u_n < \infty$ in a neighborhood of a. Thus, $A = \{z; \ u(z) = \infty\}$ is open and closed, i.e., Ω or $\emptyset$. If $u_n \nearrow u < \infty$, then u satisfies the Poisson formula by monotone convergence, and hence it is harmonic. However, if continuous functions u_n tend to a continuous limit u monotonically, then the convergence must be u.c. (this is sometimes called Dini's lemma). In the present situation one also can see that the convergence is u.c. from the representation by the Poisson integral. □

1.8 Schwarz' Reflection Principle. *Suppose that Ω is a connected domain, symmetric with respect to the real axis, and that $L = \Omega \cap \mathbb{R}$ is an interval. Let $\Omega^+ = \{z \in \Omega; \ \text{Im } z > 0\}$. Suppose that $f \in A(\Omega^+)$ and that $\text{Im } f$ has a continuous extension to $\Omega^+ \cup L$ that vanishes on L. Then there is a $F \in A(\Omega)$ such that $F = f$ in Ω^+ and $F(z) = \overline{f(\bar{z})}$ in $\Omega \setminus \Omega^+$.*

Proof. If $v = \text{Im } f$ is extended to $\Omega \setminus \Omega^+$ by letting $v(\bar{z}) = -v(z)$, then $v \in C(\Omega)$ and has the mean value property at each point; therefore, it is harmonic by Proposition 1.5. In a simply connected symmetric neighborhood ω of L in Ω there is a $g \in A(\omega)$ such that $v = \text{Im } g$ and $f = g$ in

$\Omega^+ \cap \omega$ (why?). Since g is real on L, its power series at points on L has real coefficients, and therefore $g(\bar{z}) = \overline{g(z)}$. Let $F(z) = \overline{f(\bar{z})}$ in $\Omega \setminus (\Omega^+ \cup L)$ and $F = f$ in Ω^+. Then F is analytic in $\Omega \setminus L$, but $F = g$ in $\omega \setminus L$, and hence F extends to an analytic function in Ω.   □

Exercise 2. Let Ω be as above. Suppose that $f \in A(\Omega^+)$ and that $|f| \to 1$ on L. Formulate the corresponding reflection theorem. Then assume instead that Ω is an appropriate neighborhood of some connected subset L of T and $\Omega^+ = U \cap \Omega$. Formulate the reflection theorem in this case. Hint: Use a conformal mapping from the upper half-plane onto U.

§2. Subharmonic Functions

Subharmonicity is a fundamental concept in complex analysis. Many important properties of analytic functions depend simply on the fact that the logarithms of their modulii are subharmonic. Subharmonic functions are related to harmonic functions in much the same way as convex functions (in one or several real variables) are related to linear functions.

Definition. A function u in Ω with values in $[-\infty, \infty)$ is subharmonic, $u \in SH(\Omega)$, if
(a) u is upper semicontinuous.
(b) for each $K \subset \Omega$ and $h \in C(K)$ that is harmonic in $int(K)$ and $\geq u$ on ∂K one has that $h \geq u$ on K.

Sometimes one excludes $u \equiv -\infty$ from the definition. It is clear that real harmonic functions are subharmonic. Note that if u is upper semicontinuous on a compact set K, then $\sup_K u$ is finite and attained at some point on K. Hence, if $u \in SH(\Omega)$ and $K \subset \Omega$, then $\sup_K u$ is attained on ∂K (why?).

2.1 Theorem. *Suppose that u is upper semicontinuous in Ω with values in $[-\infty, \infty)$. Then the following conditions are equivalent:*
(i) u is subharmonic.
(ii) If $D = D(z, r) \subset\subset \Omega$, $h \in C(\overline{D})$ is harmonic in D and $\geq u$ on ∂D, then $h \geq u$ in D.
(iii) If $D = D(z, r) \subset\subset \Omega$, then

$$u(z) \leq \frac{1}{2\pi} \int_0^{2\pi} u(z + re^{i\theta}) d\theta. \tag{2.1}$$

(iv) For each $z \in \Omega$, (2.1) holds for small $r > 0$.

As u is bounded from above on compact sets, the integral in (2.1) makes sense and takes values in $[-\infty, \infty)$. From condition (iv) it follows that subharmonicity is a local property. Since a harmonic function u has the mean value property, it follows that $|u|$ satisfies (iii) and hence that $|u|$ is subharmonic. If f is analytic in Ω, then $\log|f|$ is upper semicontinuous, takes values in $[-\infty, \infty)$, and satisfies condition (iv) as it is harmonic outside the zeros of f. We therefore obtain

2.2 Corollary. *If f is analytic in Ω, then $u = \log|f|$ is subharmonic in Ω.*

Proof of Theorem 2.1. Trivially (i) implies (ii). Assume that (ii) holds and take $D(z,r) \subset\subset \Omega$. Since u is upper semicontinuous, there are continuous functions h_j such that $h_j \searrow u$ on ∂D (prove this!). Let H_j be the harmonic extensions of h_j to D. Then

$$u(z) \leq H_j(z) = \frac{1}{2\pi} \int_0^{2\pi} h_j(z + re^{i\theta})d\theta,$$

according to (ii). Now (iii) follows by monotone convergence.

Clearly, (iii) implies (iv). Assume that (iv) holds and let $K \subset \Omega$, $h \in C(K)$, $h \geq u$ on ∂K, and h be harmonic in $\mathrm{int}(K)$. Then $v = u - h \leq 0$ on ∂K and satisfies (2.1) at interior points for small r. As v is upper semicontinuous, the same argument as in the proof of Theorem 1.5 gives that $v \leq 0$ on K.   □

2.3 Proposition. *If ϕ is increasing and convex on $\mathbb{R}$ and $u \in SH(\Omega)$, then $\phi \circ u \in SH(\Omega)$ (where $(\phi(-\infty)) := \lim_{x \to -\infty} \phi(x)$).*

Proof. First notice that $\phi \circ u$ is upper semicontinuous, since ϕ is increasing and continuous (any convex function on $\mathbb{R}$ is continuous). Moreover, for small r,

$$\phi \circ u(z) \leq \phi\left(\frac{1}{2\pi} \int_0^{2\pi} u(z + re^{i\theta})d\theta\right) \leq \frac{1}{2\pi} \int_0^{2\pi} \phi \circ u(z + re^{i\theta})d\theta,$$

where the first inequality holds because ϕ is increasing and the second one follows from Jensen's inequality. Now the proposition follows from Theorem 2.1.   □

2.4 Example. If $f \in A(\Omega)$, then not only is $\log|f|$ subharmonic but, e.g., $\log^+|f| := \max(\log|f|, 0)$ and $|f|^a$, $a > 0$ also are. This follows immediately from Proposition 2.3 by taking $\phi(x) = \max(x, 0)$ and $\phi(x) = \exp(ax)$, respectively.

2.5 Proposition.
(a) If $u_\alpha \in SH(\Omega)$, $u = \sup_\alpha u_\alpha < \infty$, and u is upper semicontinuous (e.g., if $\{\alpha\}$ is finite), then $u \in SH(\Omega)$.
(b) If $u_j \in SH(\Omega)$, then $u_1 + u_2 \in SH(\Omega)$ and $cu_1 \in SH(\Omega)$ if $c \geq 0$.
(c) If $u_j \searrow u$ and $u_j \in SH(\Omega)$, then $u \in SH(\Omega)$.

Proof. (a) and (b) immediately follow (e.g., from Theorem 2.1). For (c), first note that $\{u < \alpha\} = \cup_j \{u_j < \alpha\}$ is open and therefore u is upper semicontinuous. Then (c) follows by monotone convergence (and Theorem 2.1). $\square$

2.6 Theorem. *If $u \in SH(\Omega)$ and $D(z, R) \subset \Omega$, then*

$$r \mapsto m(r) = \frac{1}{2\pi} \int_0^{2\pi} u(z + re^{i\theta})d\theta, \quad 0 < r < R$$

is increasing and $m(r) \searrow u(z)$ when $r \searrow 0$.

In fact, $m(r)$ is continuous for $r < R$; see Exercise 32.

Proof. Take $r_1 < r_2$ and continuous $h_j \searrow u$ on $\partial D(z, r_2)$, and let H_j be the harmonic extensions. Then $u \leq H_j$ on $D(z, r_1)$ and hence

$$m(r_1) \leq \frac{1}{2\pi} \int_0^{2\pi} H_j(z + r_1 e^{i\theta})d\theta = H_j(z) = \frac{1}{2\pi} \int_0^{2\pi} h_j(z + r_2 e^{i\theta})d\theta;$$

but the right-hand side tends to $m(r_2)$ by monotone convergence, and thus $r \mapsto m(r)$ is increasing. If $u(z) < \alpha$, then $u(z + re^{i\theta}) < \alpha$ for small $r > 0$ because of the semicontinuity, and thus $\limsup_{r \searrow 0} m(r) \leq u(z)$. Since $u(z) \leq m(r)$, the theorem follows. $\square$

2.7 Theorem. *If Ω is connected, $u \in SH(\Omega)$, and $u \not\equiv -\infty$, then $u \in L^1_{\text{loc}}(\Omega)$; in particular, $u > -\infty$ a.e.*

Proof. From (2.1) we get that

$$u(z) \leq \frac{1}{\pi r^2} \int_{|\zeta| < r} u(z + \zeta)d\lambda(\zeta)$$

if $D(z, r) \subset\subset \Omega$. If $u(z) > -\infty$, it thus follows that u is integrable in D since u is bounded from above on $\overline{D}$. Let

$$E = \{z \in \Omega; \ u \text{ is integrable in a neighborhood of } z\}.$$

Then trivially E is open. From the argument above it follows that $u \equiv -\infty$ in a neighborhood of each point in $\Omega \setminus E$, and thus E is closed. $\square$

2.8 Corollary. *If Ω is connected and $u \not\equiv -\infty$, then the integrals in (2.1) are $> -\infty$ for all $r > 0$.*

Our next objective is to approximate an arbitrary subharmonic function with smooth ones. Let $u \in SH(\Omega)$, $u \not\equiv -\infty$, and choose a positive $\psi \in C_0^\infty(U)$ such that $\psi(z) = \psi(|z|)$ and $\int \psi d\lambda = 1$. If $\psi_\epsilon(z) = \epsilon^{-2}\psi(z/\epsilon)$, then

$\int \psi_\epsilon d\lambda = 1$ and

$$u_\epsilon = u * \psi_\epsilon = \int u(z - \epsilon\zeta)\psi(\zeta)d\lambda(\zeta)$$

is in $C^\infty(\Omega_\epsilon)$, where $\Omega_\epsilon = \{z \in \Omega; \ d(z, \partial\Omega) > \epsilon\}$.

2.9 Proposition. $u_\epsilon \in SH(\Omega_\epsilon) \cap C^\infty(\Omega_\epsilon)$ and $u_\epsilon \searrow u$ when $\epsilon \searrow 0$.

Proof. Take an arbitrary point in Ω. With no loss of generality we may assume that it is 0. Then by Fubini's theorem and (2.1),

$$u_\epsilon(0) = \int u(-\epsilon\zeta)\psi(\zeta)d\lambda(\zeta) \le \int \frac{1}{2\pi}\int_0^{2\pi} u(re^{i\theta} - \epsilon\zeta)d\theta \, \psi(\zeta)d\lambda(\zeta)$$

$$= \frac{1}{2\pi}\int_0^{2\pi} u_\epsilon(re^{i\theta})d\theta,$$

so (2.1) is satisfied for u_ϵ, and hence it is subharmonic. Now

$$u_\epsilon(0) = \int_0^\infty \int_0^{2\pi} u(-\epsilon re^{i\theta})d\theta r\psi(r)dr,$$

which decreases when ϵ decreases and has limit $\int_0^\infty 2\pi u(0)r\psi(r)dr = u(0)$ by monotone convergence and Theorem 2.6, i.e., $u_\epsilon \searrow u$ when $\epsilon \searrow 0$. $\square$

In particular, if $u, v \in SH(\Omega)$ and $u = v$ a.e., then $u \equiv v$.

2.10 Lemma. If $u \in C^2(\Omega)$, then u is subharmonic if and only if $\Delta u \ge 0$ in Ω.

Proof. Suppose that $\Delta u \ge 0$. Take a disk $D \subset\subset \Omega$, an h harmonic in a neighborhood of $\overline{D}$, and $\ge u$ on ∂D. For simplicity, suppose that $D = D(0, r)$. If $\epsilon > 0$, then $v = u - h + \epsilon|z|^2 - \epsilon r^2$ is a C^2 function such that $\Delta v = \Delta u + 4\epsilon > 0$ and $v \le 0$ on ∂D. If $\sup_D v > 0$, then v has a local maximum at some $p \in D$, and thus the Hessian at p must be negatively semidefinite, which contradicts the fact that $\Delta v =$ the trace of the Hessian > 0. Thus, $v \le 0$ in D, i.e., $u - h \le \epsilon r^2$ in D, and since ϵ is arbitrary, then $u \le h$ in D. This proves that u is subharmonic, according to Theorem 2.1.

Conversely, assume that u is subharmonic and take a point, say 0, in Ω. By Taylor's formula,

$$u(re^{i\theta}) = u(0) + 2\operatorname{Re} u_z(0)re^{i\theta} + \operatorname{Re} u_{zz}(0)r^2 e^{2i\theta} + u_{z\bar{z}}(0)r^2 + o(r^2).$$

Integrating this equality and using (2.1) we get

$$0 \le u_{z\bar{z}}(0)r^2 + o(r^2)$$

for small r, which implies that $\Delta u(0) = 4u_{z\bar{z}}(0) \ge 0$. $\square$

2.11 Theorem. *If u is subharmonic in Ω (and $\not\equiv -\infty$ on each component), then*

$$\int u\Delta\phi d\lambda \geq 0, \quad 0 \leq \phi \in C_0^2(\Omega). \tag{2.2}$$

Thus, $\phi \mapsto \int u\Delta\phi d\lambda$ is a positive linear functional on $C_0^2(\Omega)$, and hence it is represented by a positive measure (see Exercise 28) that we denote by Δu or $\Delta u d\lambda$.

Proof. Take $u_\epsilon \searrow u$, which are C^∞ and subharmonic. Then

$$\int u\Delta\phi = \lim \int u_\epsilon \Delta\phi = \lim \int \phi\Delta u_\epsilon \geq 0.$$

$\square$

2.12 Theorem. *If $u \in L^1_{\text{loc}}(\Omega)$ (or $\mathcal{D}'(\Omega)$) and (2.2) holds, then there is a unique $v \in SH(\Omega)$ such that $u = v$ a.e. (or in $\mathcal{D}'(\Omega)$).*

In particular, this implies Weyl's lemma for harmonic functions.

Proof. Let $u_\epsilon = u * \phi_\epsilon$ as before. Since $\Delta u_\epsilon = u * \Delta\phi_\epsilon \geq 0$ by (2.2), u_ϵ are subharmonic by Lemma 2.10. However, then $u * \phi_\epsilon * \phi_\delta \leq u * \phi_\epsilon * \phi_{\delta'}$ if $\delta < \delta'$, i.e., $u * \phi_\delta * \phi_\epsilon \leq u * \phi_{\delta'} * \phi_\epsilon$. Letting $\epsilon \searrow 0$, we find that $u * \phi_\delta$ is subharmonic and decreases if δ decreases. By Proposition 2.5 (c), it follows that $v = \lim_{\delta \searrow 0} u * \phi_\delta$ is subharmonic, and since $u * \phi_\delta \to u$ in L^1_{loc} (see Exercise 27), we can conclude that $u = v$ a.e. $\square$

Thus, roughly speaking, u is subharmonic if and only if Δu is a positive measure.

2.13 Example. Let us compute $\Delta \log |f|$ if f is analytic. Suppose that $f \in A(\Omega)$ has zeros at $a_1, a_2, \ldots$ with multiplicities $m_1, m_2, \ldots$. As $\Delta \log |z| = 2\pi\delta$ (see 1.7 (c) in Ch. 1 or the proof of Jensen's formula below), it follows that

$$\Delta \log |f| = 2\pi \sum m_j \delta_{a_j},$$

where δ_{α_j} is the point mass at α_j.

Let us consider the inhomogenous Laplace equation. If $f \in C^\infty(\Omega)$, then there is a solution to

$$\Delta u = f \tag{2.3}$$

in $C^\infty(\Omega)$. This follows from Theorem 2.3 in Ch. 3 applied twice. If f behaves reasonably well near the boundary, then the *Newton potential* of

f,

$$u(z) = \frac{1}{2\pi} \int_\Omega \log|z - \zeta| f(\zeta) d\lambda(\zeta), \tag{2.4}$$

is a solution in Ω. In general, one is interested in a solution with prescribed boundary values. If v is a solution to (2.3) in U, which is continuous up to the boundary, then certainly $u = v - Pv$ is the unique solution that vanishes on the boundary.

2.14 The Green Potential. For $\zeta, z \in U$ let

$$G(\zeta, z) = \frac{1}{2\pi} \log\left|\frac{\zeta - z}{1 - \bar\zeta z}\right|.$$

For fixed $z \in U$, $G(\zeta, z) = \mathcal{O}(1 - |\zeta|)$; thus, if μ is a positive measure in U such that

$$\int_U (1 - |\zeta|) d\mu(\zeta) < \infty, \tag{2.5}$$

we can define *the Green potential*

$$\mathcal{G}\mu(z) = \int_U G(\zeta, z) d\mu(\zeta).$$

We leave it to the reader to verify that if μ is any positive measure that satisfies (2.5), then $\mathcal{G}\mu(z)$ is a negative subharmonic function in U that solves $\Delta u = \mu$ in the distribution sense and vanishes on the boundary in the sense that

$$\lim_{r \nearrow 1} \frac{1}{2\pi} \int_0^{2\pi} \mathcal{G}\mu(re^{i\theta}) d\theta = 0.$$

If $\mu = f$ is a bounded function, then $\mathcal{G}f$ has a continuous extension to $\overline{U}$, and hence it vanishes on the boundary and is therefore the unique solution to (2.3) in U that vanishes on the boundary. Consequently, for any v in, for example, $C^2(\overline{U})$, we have *the Riesz decomposition*

$$v(z) = \mathcal{G}(\Delta v)(z) + Pv(z). \tag{2.6}$$

Later on it will turn out that this decomposition holds for all subharmonic functions satisfying (2.8).

We conclude this section with Jensen's formula, which expresses the connection between the growth of the mean values $m(r)$ and the size of Δu for a subharmonic function u.

2.15 Jensen's Formula. *If* $u \in SH(D(a, R))$ *and* $u \not\equiv -\infty$, *then*

$$
\begin{aligned}
u(a) = &\frac{1}{2\pi} \int_0^{2\pi} u(a + re^{i\theta}) d\theta \\
&- \frac{1}{2\pi} \int_{|z-a|<r} \log \frac{r}{|z-a|} \Delta u(z) d\lambda(z), \quad r < R.
\end{aligned}
\tag{2.7}
$$

Proof. Let $a = 0$ and assume first that u is C^2. Green's identity applied to the domain $\{\epsilon < |z| < r\}$ yields

$$
\begin{aligned}
-\int_{\epsilon < |z| < r} &\log \frac{r}{|z|} \Delta u(z) d\lambda(z) \\
&= -\int_0^{2\pi} u(re^{i\theta}) d\theta + \int_0^{2\pi} u(\epsilon e^{i\theta}) d\theta + \mathcal{O}(\epsilon \log \epsilon),
\end{aligned}
$$

which gives (2.7) when $\epsilon \searrow 0$. For a general subharmonic u, let $u_\epsilon = u * \psi_\epsilon$ as before. Since (2.7) holds for each u_ϵ, we just have to ensure that we can take limits. If $g(z) = \log(r/|z|)$ is defined as zero outside $\{|z| < r\}$, then g is continuous except at the origin. Now observe that

$$
\int g(z) \Delta u_\epsilon(z) = \int g(z) \psi_\epsilon * \Delta u(z) = \int g * \psi_\epsilon(z) \Delta u(z).
$$

However, $g * \psi_\epsilon \to g$ uniformly outside a neighborhood of the origin (see Exercise 27) and $g * \psi_\epsilon \nearrow g$ in a neighborhood of the origin (since $-g$ is subharmonic there); therefore, $\int g \Delta u_\epsilon \to \int g \Delta u$. Since the convergence of the other terms in (2.7) is obvious, Jensen's formula is proved. $\qquad \square$

2.16 Corollary. *If* $u \in SH(U)$ *and*

$$
\sup_{r<1} \int_0^{2\pi} u^+(re^{i\theta}) d\theta < \infty,
\tag{2.8}
$$

then $\mu = \Delta u$ *satisfies* (2.5).

We leave the verification of this as an exercise. Conversely, if μ satisfies (2.5), then $\mathcal{G}\mu$ is a solution to (2.3) such that (2.8) holds; cf. 2.14.

Supplementary Exercises

Exercise 3. Suppose that $u \in C^2(\Omega)$. Show directly that $\Delta u = 0$ if (1.1) holds for small $r > 0$.

Exercise 4. Suppose that Ω is connected. Show that if each real harmonic u is the real part of some analytic function in Ω, then Ω must be simply connected. (Together with Proposition 1.1, this establishes the equivalence of condition (k) to the other conditions in Theorem 4.1 in Ch. 3.)

Exercise 5. Assume that f and f^2 are harmonic (and Ω is connected). Show that f or $\bar{f}$ is analytic.

Exercise 6. Suppose that u and v are real and harmonic. If uv is also harmonic, what is the relation between u and v?

Exercise 7. What can be said about $\{\nabla u = 0\}$ if u is harmonic?

Exercise 8. Show that $u \circ f$ is harmonic if u is harmonic and f is analytic.

Exercise 9. Suppose that $u \in L^1_{loc}(\Omega)$ and

$$\frac{1}{|D|}\int_D u = u(a)$$

for all $D = D(a,r) \subset\subset \Omega$. Show that u is harmonic.

Exercise 10. Suppose that $f_j \in A(\Omega)$, Ω is connected, $f_j(a)$ converges for some $a \in \Omega$, and $\operatorname{Re} f_j$ converges u.c. Show that f_j converges u.c.

Exercise 11. Suppose that $u_j \to u$ in $L^1_{loc}(\Omega)$ and u_j are harmonic. Show that $u_j \to u$ u.c.

Exercise 12. Prove the maximum principle for harmonic functions by using only the fact that they satisfy $\Delta u = 0$.

Exercise 13. Show that if u is real, harmonic in $\mathbb{C}$, and bounded from above, then it is constant.

Exercise 14. Suppose that u is real and harmonic in $\{0 < |z| < 2\}$, and $u(z) = o(\log(1/|z|))$ when $z \to 0$. Show that u has a removable singularity at 0. Hint: Let v be the harmonic function in U that equals u on the unit circle. Note that for any $\epsilon > 0$, $u - v - \epsilon \log|z| \le 0$ on $\partial\{\delta < |z| < 1\}$ if δ is small enough.

Exercise 15. Show that it is enough to verify (ii) in Theorem 2.1 for all h that are harmonic in some neighborhood of $\bar{D}$.

Exercise 16. Suppose that f is analytic and nonconstant. Use the preceding exercise to show that $\log|f|$ is subharmonic. Hint: Suppose that g is analytic in a neighborhood of $D \subset\subset \Omega$ and that $\log|f| \le \operatorname{Re} g$ on ∂D. Then the maximum principle applied to $f \exp(-g)$ implies that the inequality holds in D as well.

Exercise 17. A function $u \in C^2(\Omega)$, $\Omega \subset \mathbb{R}^n$ is *harmonic* if $\Delta u = \sum \partial^2 u / \partial x_j^2 = 0$. Prove the mean value property and the maximum principle for such functions.

Exercise 18. A continuous real function u in $\Omega \subset \mathbb{R}^n$ is *convex* if $t \mapsto u(x + ty)$ is convex for all $x \in \Omega$, $y \in \mathbb{R}^n$, and small real t. Show that a

$u \in C^2(\Omega)$ is convex if and only if the Hessian $(\partial^2 u/\partial x_j \partial x_k)_{jk}$ is positively semidefinite for each $x \in \Omega$.

Exercise 19. Show that u is subharmonic if it is convex. Give a counterexample for the converse.

Exercise 20. Show that if $u, v \geq 0$ and $\log u$ and $\log v$ are subharmonic, then $\log(u + v)$ is subharmonic.

Exercise 21. Supply proofs for the statements about the Green function in 2.14.
(a) Suppose that $f \in C_0^2(\mathbb{C})$. Show that $\Delta \int \log |\zeta - z| f(\zeta) d\lambda(\zeta) = 2\pi f$.
(b) Show that $\mathcal{G}\mu(z) \leq 0$ and is locally integrable in U.
(c) Show that $\mathcal{G}\mu(z)$ is subharmonic.
(d) Show that $\Delta \mathcal{G}\mu = \mu$.
(e) Show that $(1/2\pi) \int_0^{2\pi} \mathcal{G}\mu(re^{i\theta}) d\theta \nearrow 0$ when $r \nearrow 1$.
(f) Show that $\mathcal{G}f(z)$ is continuous up to the boundary if f is bounded.
Hints: For (c), one can consider $\mathcal{G}_R\mu(z)$ obtained from the kernel $G_R(\zeta, z) = \max(G(\zeta, z), -R)$. For (f), one can observe that $\mathcal{G}f$ has bounded derivatives in the interior.

Exercise 22. Suppose that $u \in SH(U)$ is not identically $-\infty$. Show that

$$\sup_{r<1} \int u^+(re^{i\theta}) d\theta < \infty \text{ iff } \sup_{1/2<r<1} \int |u(re^{i\theta})| d\theta < \infty.$$

Exercise 23. Prove Corollary 2.16!

Exercise 24. Formulate Jensen's formula when $u = \log|f|$, where f is analytic.

Exercise 25. Suppose that $u \in SH(U)$ and that $\limsup u(z_n) \leq 0$ for any sequence z_n that converges to a boundary point. Show that $u \leq 0$ in U.

Exercise 26. Suppose that u is upper semicontinuous on T. Show that there are continuous u_n such that $u_n \searrow u$.

Exercise 27. Take $\psi \in C_0^\infty(U)$ such that $\int \psi d\lambda = 1$. Let $\psi_\epsilon(z) = \epsilon^{-2}\psi(z/\epsilon)$. Show that $\psi_\epsilon * f \to f$ in L_{loc}^1 if $f \in L_{loc}^1$ and that $\psi_\epsilon * f \to f$ u.c. if f is continuous.

Exercise 28. Show that if Λ is a positive linear functional on $C_0^2(\Omega)$, then there is a positive measure μ in Ω such that $\Lambda\phi = \int \phi d\mu$ for all $\phi \in C_0^2(\Omega)$.

Exercise 29. Prove Weyl's lemma 1.6; cf. (the proof of) Theorem 2.12.

Exercise 30. Show that $u(z) = \log(1/d(z, \Omega^c))$ is subharmonic in Ω.

Exercise 31. Find all subharmonic u in U such that $u(z) = u(|z|)$.

Exercise 32. Show that $m(r) = \int u(re^{i\theta}) d\theta$ is continuous if u is subharmonic. Hint: First suppose that $u(z) = u(|z|)$.

Exercise 33. Let Ω be a doubly connected domain with smooth boundary, and let us assume that the Dirichlet's problem can be solved with continuous boundary data (which is true). Then there is a harmonic $u(z)$ such that $u = 0$ on the boundary of the bounded component and $R > 1$ on the other one. Show that if R is appropriately chosen, then u has a harmonic conjugate v in Ω that is well-defined modulo an integer multiple of 2π. Then $h(z) = \exp(u(z) + iv(z))$ is a well-defined analytic function in Ω whose image is contained in the annulus $A(1, \exp R)$. Show that $h(z)$ is in fact a conformal equivalence. Hint: First show that h is proper, i.e., $h^{-1}(K)$ is compact for each compact K. Then apply the argument principle to cycles in Ω approximating $\partial\Omega$.

Definition. A set $E \subset \mathbb{C}$ is *polar* if there is a subharmonic $u \not\equiv -\infty$ such that $E \subset \{u = -\infty\}$.

Exercise 34. Show that each countable set is polar. Also show that any polar set has measure zero, and supply a counterexample for the converse.

Exercise 35. Suppose that $u_n \in SH(\Omega)$, $\sup_n u_n \le M$, where M is locally bounded in Ω, and $\limsup_{n\to\infty} u_n \le 0$ at each point in Ω. Show that for each compact $K \subset \Omega$ and $\epsilon > 0$ there is an N such that $u_n < \epsilon$ on K for $n > N$.

Exercise 36. Show the *three-circle theorem*: If $f \in A(A(1, R))$, then

$$\log|f(z)| \le \frac{\log b/|z|}{\log b/a} \log m(a) + \frac{\log|z|/a}{\log b/a} \log m(b),$$

if $1 < a < |z| < b < R$ and $m(r) = \sup_{|z|=r} |f(z)|$.

Notes

In $\mathbb{R}^2$, harmonic functions can be considered either as real parts (locally) of analytic functions, or as homogeneous solutions to a real elliptic equation (the Laplace equation). I have tried to shed light on both aspects, and consequently some facts are proved (at least implicitly) twice. For instance, the maximum principle is derived from the mean value property (1.1) (which in turn follows from the mean value property for analytic functions, as well as from Jensen's formula (2.7)), and also directly from the Laplace equation (see Lemma 2.10).

There is a close connection between Dirichlet's problem (DP) and the Riemann mapping theorem. For instance, the latter implies solvability of DP in all simply connected domains (with reasonable boundary). In Exercise 33, it is suggested how one can obtain a conformal equivalence from a solution to DP. Solvability of DP under weak assumptions of the regularity of the boundary is studied in Potential theory, in which the

Newton and Green potentials, polar sets, and *capacity* are basic concepts; see [Ra] for a nice introduction.

The family of functions $(\psi_\epsilon)_{\epsilon>0}$ in Proposition 2.9 and Exercise 27 is called an approximate identity (in $\mathbb{R}^2$). The Poisson kernel $P_r(t)$ is an approximate identity on the circle T. Previously we have met approximate identities in Section 6 and Exercise 23 in Ch. 3.

5

Zeros, Growth, and Value Distribution

§1. Weierstrass' Theorem

Our first aim in this chapter is to prove that any subset of $\Omega \subset \mathbb{C}$ that has no limit point in Ω is the zero set of some $f \in A(\Omega)$ (Weierstrass' theorem). To this end we have to consider infinite products.

1.1 Lemma. *If z_n are complex numbers, then*

$$\left| \prod_1^N (1 + z_n) - 1 \right| \le \prod_1^N (1 + |z_n|) - 1.$$

Proof. Since

$$\prod_1^{k+1} (1 + z_n) - 1 = (1 + z_{k+1}) \left(\prod_1^k (1 + z_n) - 1 \right) + z_{k+1},$$

it follows by induction over k that

$$\left| \prod_1^{k+1} (1 + z_n) - 1 \right| \le (1 + |z_{k+1}|) \left(\prod_1^k (1 + |z_n|) - 1 \right) + |z_{k+1}|$$

$$= \prod_1^{k+1} (1 + |z_n|) - 1.$$

$\square$

1.2 Proposition. *If $f_n \in A(\Omega)$, $f_n \not\equiv -\infty$ on each component of Ω, and $\sum |1 - f_n|$ converges u.c., then*

$$f(z) = \prod_1^{\infty} f_n(z) = \lim \prod_1^N f_n(z)$$

converges u.c. in Ω. Moreover, the zeros of f are the union of the zeros of f_n, counted with multiplicities.

Proof. Take fixed $K \subset \Omega$. Note that

$$\left| \prod_1^N f_n \right| \leq \prod_1^N (1 + |1 - f_n|) \leq \exp \sum_1^N |1 - f_n| \leq C \qquad (1.1)$$

on K, independently of N (where we use that $1 + x \leq \exp x$). If $N \geq M$, then by the lemma ($z_n = f_n - 1$) and (1.1),

$$\left| \prod_1^N f_n - \prod_1^M f_n \right| = \left| \prod_1^M f_n \right| \left| \prod_{M+1}^N f_n - 1 \right|$$

$$\leq C \left(\prod_{M+1}^N (1 + |1 - f_n|) - 1 \right) \leq C \left(\exp \sum_{M+1}^N |1 - f_n| - 1 \right),$$

which tends to 0 uniformly on K when $N \geq M \to \infty$; therefore, the product converges uniformly on K. Thus, the first statement is proved.

For the second statement, note that each $z \in \Omega$ has a neighborhood where at most a finite number of f_n have zeros, since locally $\sum_{M+1}^\infty |1 - f_n| < 1/2$ if M is large enough. Now $\prod_1^\infty f_n = \prod_1^M f_n \prod_{M+1}^\infty f_n$ and

$$\left| \prod_{M+1}^N f_n - 1 \right| \leq \exp \sum_{M+1}^N |1 - f_n| - 1 \leq \exp 1/2 - 1 < 1,$$

and therefore $\prod_{M+1}^\infty f_n$ has no zeros near z. $\qquad \square$

Let

$$W_0 = 1 - z, \quad W_p(z) = (1 - z) \exp \sum_1^p \frac{z^j}{j}, \quad p \geq 1.$$

1.3 Lemma. $W_p(z) = 0$ if and only if $z = 1$; moreover, $|1 - W_p(z)| \leq |z|^{p+1}$ in U.

Proof. If $h = 1 - W_p$, then $h(0) = 0$ and $h'(z) = z^p \exp \sum_1^p z^j/j$. Hence, $\phi(z) = h(z)/z^{p+1}$ is analytic and, since ϕ has positive Taylor coefficients, $|\phi(z)| \leq \phi(|z|) \leq \phi(1) \leq 1$ for $|z| < 1$. $\qquad \square$

We now are prepared to construct functions with prescribed zeros.

1.4 Weierstrass' Theorem. *Suppose that* $\{a_n\}$ *is a sequence in* $\Omega \subset \mathbb{P}$ *with no limit point in* Ω, $\Omega \neq \mathbb{P}$, *and* m_n *is a sequence of positive integers. Then there is an* $f \in A(\Omega)$ *that has zeros of orders* m_n *at* a_n *and no others.*

Proof. We may assume that $\infty \in \Omega$ and that $\infty \neq a_n$ for each n. Let a_n be an enumeration such that each a_n occurs precisely m_n times. Take $\beta_n \in \mathbb{P} \setminus \Omega$ such that $|\beta_n - \alpha_n| = d(\alpha_n, \partial\Omega)$. We claim that

$$f(z) = \prod_1^\infty W_n \left(\frac{\alpha_n - \beta_n}{z - \beta_n} \right)$$

has the required properties. In fact, since ∞ is not a limit point of $\{\alpha_n\}$, $|\beta_n - \alpha_n| = d(\alpha_n, \partial\Omega) \to 0$ when $n \to \infty$, and hence there is for a given $K \subset \Omega$ an N such that $|z - \beta_n| \geq d(K, \partial\Omega) > 2|\alpha_n - \beta_n|$ if $n \geq N$ and $z \in K$. Thus, $|\alpha_n - \beta_n|/|z - \beta_n| \leq 1/2$ if $n \geq N$ and hence, by Lemma 1.3,

$$\left| 1 - W_n \left(\frac{\alpha_n - \beta_n}{z - \beta_n} \right) \right| \leq 2^{-n}, \quad n \geq N, z \in K.$$

It now follows from Proposition 1.2 that f has the desired properties. $\square$

Certainly this theorem, as well as the next two corollaries, fail if $\Omega = \mathbb{P}$.

1.5 Corollary. *If g is meromorphic in $\Omega \subset \mathbb{C}$, then $g = f/h$ for some $f, h \in A(\Omega)$.*

Proof. Take an h that has zeros where g has poles, and with the same multiplicities. Then $f = gh$ is analytic. $\square$

1.6 Corollary. *If α_j is a sequence with no limit point in $\Omega \subset \mathbb{C}$, f_j is analytic near α_j, and m_j are positive integers, then there is an $f \in A(\Omega)$ such that $f(z) - f_j(z) = \mathcal{O}(|z - \alpha_j|^{m_j+1})$ when $z \to \alpha_j$.*

Thus, the Taylor expansion of f can be prescribed up to order m_j at each α_j.

Proof. Take $g \in A(\Omega)$, which has zeros of orders $m_j + 1$ at α_j. According to Mittag–Leffler's theorem, there is a meromorphic h in Ω having poles only at α_j such that $h - f_j/g$ is analytic near α_j. Then let $f = gh$. $\square$

1.7 Corollary. *There is an $f \in A(\Omega)$ that cannot be continued to any larger domain. More precisely: If $a \in \Omega$, $g \in A(D(a,r))$, and $f = g$ near a, then $D(a,r) \subset \Omega$.*

Notice that for a general domain the statement about f is stronger than: If $a \in \Omega$, $g \in A(D(a,r))$ and $f = g$ in $D(a,r) \cap \Omega$, then $D(a,r) \subset \Omega$. For instance, if $\Omega = \mathbb{C} \setminus \{z; \operatorname{Im} z = 0, \operatorname{Re} z \geq 0\}$, then $\log z$ in Ω satisfies the latter statement but not the one in the corollary.

Proof of Corollary 1.7. Let α_n be an enumeration of all points in Ω with rational coordinates such that each such point occurs an infinite number of times. Let K_n be an increasing exhausting sequence of compact sets in Ω, and choose for each n a $w_n \in \Omega \setminus K_n$ such that $|a_n - w_n| \leq d(a_n, \partial\Omega)$. Since any fixed compact set is contained in some K_n, the sequence w_n has no limit point in Ω, and hence there is an analytic function f with zeros w_n and no others. Now, if $a \in \Omega$ has rational coordinates, then the largest open disk $D(a, r)$ included in Ω will contain infinitely many points w_n, and hence no function that equals f near a can be analytic (or even meromorphic) in any larger disk with center a. From this the corollary follows.   □

§2. Zeros and Growth

So far we only have considered the existence of analytic functions with pre-scribed zeros. We now shall study the connection between the "number of zeros" and the growth. As a simple example, note that if an entire function has m zeros (counted with multiplicities), then it must grow at least as fast as $|z|^m$ when $z \to \infty$. This is an instance of a general phenomenon which, roughly speaking, means that "the more zeros f has, the faster $|f|$ must grow." Conversely, if there are not too many prescribed zeros, one can find a function with these zeros that does not grow too fast.

If f is analytic in the disk $D(0, R)$, we can measure its growth by its *characteristic function*

$$T(r, f) = \frac{1}{2\pi} \int_0^{2\pi} \log \sqrt{1 + |f(re^{i\theta})|^2} d\theta - \log \sqrt{1 + |f(0)|^2}.$$

Since $z \mapsto \log \sqrt{1 + |f(z)|^2}$ is subharmonic (see Exercise 20 in Ch. 4), $T(r, f)$ is a nonnegative increasing function for $r < R$ and $T(0, f) = 0$. Note that

$$T(r, f) = \frac{1}{2\pi} \int_0^{2\pi} \log^+ |f(re^{i\theta})| d\theta + \mathcal{O}(1)$$

when $r \nearrow R$, since $|\log \sqrt{1 + x^2} - \log^+ x|$ is less than some constant ($\log \sqrt{2}$) for all $x > 0$. In the same way, one verifies that $T(r, f - a) = T(r, f) + \mathcal{O}(1)$ for each fixed a and that $T(r, af) = T(r, f) + \mathcal{O}(1)$ if $a \neq 0$. Clearly, control of $T(r, f)$ puts some growth restriction on f, contrary to $\int \log |f(re^{i\theta})| d\theta$, which is constant in r if f is nonvanishing. If f is analytic in the disk $D(0, R)$ and $r < R$, we let $n(r, f)$ denote the number of zeros of f on the closed disk $\{|z| \leq r\}$; and if $f(0) \neq 0$, we let

$$N(r, f) = \int_0^r n(s, f) \frac{ds}{s}.$$

If $f(0) = 0$, the expression

$$N(r, f) - N(r_0, f) = \int_{r_0}^r n(s, f) \frac{ds}{s}$$

is at least well defined, and its growth when $r \nearrow R$ does not depend on the choice of r_0. If $\alpha_1, \alpha_2, \ldots$ are the zeros of f (counted with multiplicities), we claim that

$$N(r, f) = \sum_{|\alpha_j| \le r} \log \frac{r}{|\alpha_j|}. \tag{2.1}$$

In fact, if $|\alpha_1| \le |\alpha_2| \le \ldots$, we have

$$\int_0^r n(s, f) \frac{ds}{s} = \sum_1^{M-1} \int_{|\alpha_k|}^{|\alpha_{k+1}|} n(s, f) \frac{ds}{s} + \int_{|\alpha_M|}^r n(s, f) \frac{ds}{s}$$

$$= \sum_1^{M-1} k \log \frac{|\alpha_{k+1}|}{|\alpha_k|} + M \log \frac{r}{|\alpha_M|} = \sum_1^M \log \frac{r}{|\alpha_k|}.$$

A more suggestive verification can be obtained by noting that $dn/ds = \sum_1^N \delta_{|\alpha_k|}$ in the distribution sense, since (2.1) then immediately follows from a (formal) integration by parts.

2.1 Jensen's Formula. Let f be analytic in $D(0, R)$, $r < R$, and let α_n be the zeros of f on $\overline{D(0, r)}$, ordered such that $|\alpha_n| \le |\alpha_{n+1}|$. Assume that $f(0) \ne 0$. Then

$$N(r, f) = \sum \log \frac{r}{|\alpha_n|} = \frac{1}{2\pi} \int_0^{2\pi} \log |f(re^{i\theta})| d\theta - \log |f(0)|. \tag{2.2}$$

In fact, this is simply Jensen's formula in Ch. 4 applied to $u = \log |f|$. An independent proof is outlined in Exercise 3. This formula implies a restriction on the amount of zeros of f in terms of its growth. In fact, for any f that is analytic in $D(0, R)$ we get (if $f(0) = 0$ apply Jensen's formula to $f(z)/z^M$)

$$N(r, f) - N(r_0, f) \le T(r, f) + \mathcal{O}(1). \tag{2.3}$$

We now shall consider the class of functions in the unit disk with a bounded characteristic function, the *Nevanlinna class*

$$\mathcal{N} = \left\{ f \in A(U); \sup_{r<1} \frac{1}{2\pi} \int_0^{2\pi} \log^+ |f(re^{i\theta})| d\theta < \infty \right\}.$$

The inequality (2.3) implies that $N(r, f)$ is bounded for each $f \in \mathcal{N}$. This can be rephrased in the following way.

2.2 Theorem. If $\alpha_1, \alpha_2, \ldots$ are the zeros, counted with multiplicities, to an $f \in \mathcal{N}$, then they satisfy the Blaschke condition

$$\sum 1 - |\alpha_j| < \infty.$$

Proof. We may assume that $f(0) \neq 0$. The Nevanlinna condition and Jensen's formula (or (2.3)) imply that

$$C \geq \sum_{|\alpha_k| \leq r} \log \frac{r}{|\alpha_k|} \geq \sum_{|\alpha_k| \leq r} (r - |\alpha_k|),$$

for some constant C and all $r < 1$, which proves the theorem. □

We now shall see that, conversely, any sequence satisfying the Blaschke condition occurs as the zero set of some Nevanlinna function. Actually, there is such a function that is even bounded. Let H^∞ denote the space of bounded analytic functions in U with norm

$$\|f\|_{H^\infty} = \sup_U |f|.$$

Clearly, $H^\infty \subset \mathcal{N}$.

2.3 Theorem. *If $\{\alpha_n\}$ is a sequence in U such that $\sum 1 - |\alpha_n| < \infty$ and each $\alpha_n \neq 0$, then*

$$B(z) = z^k \prod_1^\infty \frac{\alpha_n - z}{1 - \bar{\alpha}_n z} \frac{|\alpha_n|}{\alpha_n} \tag{2.4}$$

is in H^∞, $\|B\|_\infty = 1$, and B has precisely the zeros α_n and 0 (counted with multiplicities). Moreover,

$$B^*(e^{i\theta}) = \lim_{r \nearrow 1} B(re^{i\theta})$$

exists for a.e. θ, $|B^*(e^{i\theta})| = 1$ for a.e. θ, and

$$\lim_{r \nearrow 1} \frac{1}{2\pi} \int_0^{2\pi} \log |B(re^{i\theta})| d\theta = 0. \tag{2.5}$$

Proof. Since

$$\left| 1 - \frac{\alpha_n - z}{1 - \bar{\alpha}_n z} \frac{|\alpha_n|}{\alpha_n} \right| = \left| \frac{(\alpha_n + |\alpha_n|z)(1 - |\alpha_n|)}{(1 - \bar{\alpha}_n z)\alpha_n} \right| \leq \frac{1+r}{1-r}(1 - |\alpha_n|), \quad |z| \leq r,$$

it is clear that the product converges u.c., that B has the desired zeros, and that $|B| \leq 1$. For the existence of the radial limits B^*, see Theorem 2.2 in Ch. 6. The limit (2.5) exists since $\log |B|$ is subharmonic and nonpositive. If B_N is the product of the N first factors in (2.4), then B_N is continuous on $\bar{U}$ and has modulus 1 on T; therefore,

$$\log |(B/B_N)(0)| \leq \lim_{r \nearrow 1} \frac{1}{2\pi} \int_0^{2\pi} \log |(B/B_N)(re^{i\theta})| d\theta$$

$$= \lim_{r \nearrow 1} \frac{1}{2\pi} \int_0^{2\pi} \log |B(re^{i\theta})| d\theta + 0 \leq \frac{1}{2\pi} \int_0^{2\pi} \log |B^*(e^{i\theta})| d\theta \leq 0,$$

where the next to last inequality follows from Fatou's lemma (keeping in mind that $\log |B| \le 0$). Since $\log |(B/B_N)(0)| \nearrow 0$ when $N \to \infty$, (2.5) and the equality $|B^*| = 1$ a.e. follow. $\Box$

A product of the form (2.4) is called a *Blaschke product*. These products will play an important role in subsequent chapters.

2.4 Remark. The problem of finding an analytic function with prescribed zeros and a certain growth can be rephrased as a problem for the inhomogeneous Laplace equation. For the sake of simplicity, let us assume that Ω is simply connected. The case when Ω is finitely connected works in essentially the same way; see Exercise 11. If $\Delta u = \sum \delta_{a_j}$ (in the distribution sense), we claim that there is an analytic f with zero set $\{a_j\}$ such that $u = \log |f|$. Thus the problem is to solve the Laplace equation with control of the growth of the solution. In fact, if $h(z)$ is an arbitrary function with the prescribed zeros, then $u(z) - \log |h(z)|$ is harmonic and hence it is $\operatorname{Re} g(z)$ for some analytic $g(z)$. Then $f(z) = h(z) \exp g(z)$ works. Notice that the Blaschke product is precisely the solution obtained in this way from the solution to $\Delta u = \sum \delta_{a_j}$ given by the Green function; cf. 2.14 in Ch. 4.

One also can construct entire functions with prescribed zeros and with control of the growth. Let α_n be a sequence (with multiplicities) in $\mathbb{C}$ with no limit point. Using the recipe in the proof of Weierstrass' theorem we get the entire function

$$f(z) = \prod_{1}^{\infty} W_{p_n}(z/\alpha_n)$$

with zeros α_n, where $p_n = n - 1$. However, if one has some control of the sequence α_n, one can choose smaller p_n. For instance, if $\sum 1/|\alpha_n|^{p+1} < \infty$, one can take $p_n = p$ and form the product $\prod_{1}^{\infty} W_p(z/\alpha_n)$, which is $\mathcal{O}(\exp |z|^{p+1})$ when $|z| \to \infty$; see Exercises 4 and 5. An entire function that satisifies such a growth condition for some p is said to be of *finite order*.

§3. Value Distribution of Entire Functions

We now turn our attention to value distribution of entire functions. Since $T(r, f) = T(r, f - a) + \mathcal{O}(1)$, Jensen's formula also puts some restrictions on how often the value a can be attained by the function f. If $n(r, f, a) = n(r, f - a)$ and $N(r, f, a) = N(r, f - a)$, then

$$N(r, f, a) - N(r_0, f, a) \le T(r, f) + \mathcal{O}(1). \tag{3.1}$$

We will show that an entire function must attain all but a few exceptional values as often as is possible in view of (3.1). In particular, the little Picard theorem will follow.

We first notice that one can get uniform estimates from estimates of the characteristic function. Without loss of generality we may assume that $f(0) = 0$; otherwise, we can replace f with $f - f(0)$. Since $u(z) = \log \sqrt{1 + |f(z)|^2}$ is subharmonic,

$$\int_{D(0,r)} u\,d\lambda \leq T(r, f)\pi r^2 \tag{3.2}$$

for all r; and since $u \geq 0$, then

$$u(z) \leq \frac{1}{\pi|z|^2} \int_{D(z,|z|)} u\,d\lambda \leq \frac{1}{\pi|z|^2} \int_{D(0,2|z|)} u\,d\lambda \leq 4T(2|z|, f) \tag{3.3}$$

for all z if $f(0) = 0$.

3.1 Lemma. *If f is entire and*

$$\liminf_{r\to\infty} \frac{T(r, f)}{\log r} \leq m,$$

then f is a polynomial of degree at most m.

Proof. First suppose that $m = 0$. Then the condition implies that $T(r_j, f) \leq \epsilon \log r_j$ for some sequence $r_j \to \infty$. Hence, by (3.3),

$$\sup_{|z|\leq r_j/2} |f(z)| \leq r_j^{4\epsilon},$$

and thus f is constant; cf. the proof of Liouville's theorem. Now one can proceed by induction, noting that $T(r, (f(z)-f(0))/z) \leq T(r, f(z)-f(0)) - \log r + \mathcal{O}(1) = T(r, f) - \log r + \mathcal{O}(1)$. $\qquad\square$

For a nonconstant entire function f, the *defect* with respect to the value a is defined as

$$\delta(f, a) = 1 - \limsup_{r\to\infty} \frac{N(r, f, a)}{T(r, f)},$$

so that the defect $\delta(f, a)$ is a number between 0 and 1 that measures to what extent f fails to attain the value a as often as it possibly can in view of (3.1).

3.2 Example. Consider $f(z) = \exp(z^k)$ for some positive integer k. Since $f(z)$ has no zeros, $N(r, f, 0) = 0$ and hence $\delta(f, 0) = 1$. On the other hand,

$$T(r, f) = \frac{1}{2\pi} \int_0^{2\pi} \log^+ \exp(r^k \cos(k\theta))d\theta + \mathcal{O}(1),$$

and therefore $\lim_{r \to \infty} T(r, f)/r^k = 1/\pi$. Since $n(r, f, 1) \sim kr^k/\pi$, we have that $N(r, f, 1) \sim r^k/\pi$ and hence $\delta(f, 1) = 0$. In the same way one can verify that $\delta(f, a) = 0$ for all $a \neq 0$.

This example illustrates a general phenomenon.

3.3 Theorem (The Defect Relation). *If f is entire and nonconstant, and $a_1, a_2, \ldots, a_k$ are distinct complex numbers, then*

$$\sum_1^k \delta(f, a_j) \leq 1.$$

In particular, the defect must be zero for all but a countable set of values. If f totally avoids some value, then the defect is zero for all other values, i.e., the little Picard theorem follows. Moreover, if f attains some value a only a finite number of times, then $N(r, f, a) = m \log r + \mathcal{O}(1)$, and therefore the defect with respect to a is 1 unless f is a polynomial of degree m (cf. Lemma 3.1). Hence, any entire f that is not a polynomial attains all values, with one possible exception, infinitely many times. If f is a polynomial, then the defect $\delta(f, a) = 0$ for all values a, and thus we have strict inequality in the defect relation.

3.4 Remark. One also can define characteristic function and defect for meromorphic functions; the corresponding result then is that for a nonconstant meromorphic f no sum of defects can exceed 2. Since an analytic function is just a meromorphic one that avoids ∞, this statement immediately implies Theorem 3.3.

The rest of this chapter is devoted to the proof of Theorem 3.3. Even though we restrict ourselves to the case when f avoids the value ∞, it is natural to adopt an invariant view of $\mathbb{P}$. If it is parametrized by $z \in \mathbb{C}$ as usual, then the measure

$$d\mu(z) = \frac{1}{\pi} \frac{d\lambda(z)}{(1 + |z|^2)^2}$$

is invariant with respect to the unitary linear fractional transformations of $\mathbb{P}$ (cf. Exercises 33–36 in Ch. 2), i.e., the measure of a given (measurable) set is not changed if it is moved by such a mapping. Moreover, we have the invariant distance

$$\chi(z, w) = \frac{|z - w|}{\sqrt{1 + |z|^2}\sqrt{1 + |w|^2}};$$

$\chi(z, w) = 0$ if and only if $z = w$, and $\chi(z, w) \leq 1$ with equality if and only if z and w are antipodal on $\mathbb{P}$.

For a meromorphic f in $D(0, R)$ and $f(0) \neq a \in \mathbb{P}$ we define the *proximity function*

$$m(r, f, a) = \frac{1}{2\pi} \int_0^{2\pi} \log \frac{1}{\chi(f(re^{i\theta}), a)} d\theta - \log \frac{1}{\chi(f(0), a)}.$$

It measures how close to a the values of f are on the circle $|z| = r$. Jensen's formula can be rephrased as (check!)

3.5 Proposition. *If f is analytic in $D(0, R)$ and $a \neq f(0)$, then*

$$T(r, f) = m(r, f, a) + N(r, f, a), \quad r < R. \tag{3.4}$$

Note that (3.4) is exact, i.e., not modulo $\mathcal{O}(1)$. By the invariance of $d\mu$ and χ it follows (Exercise 17) that $\int_{\mathbb{P}} m(r, f, a) d\mu(a) = 0$ and hence

$$\int_{\mathbb{P}} N(r, f, a) d\mu(a) = T(r, f);$$

therefore, at least $N(r, f, a)$ equals $T(r, f)$ in the mean. Moreover, it follows that the proximity function $m(r, f, a)$ is negative for some values a. Nevertheless, we have

3.6 Lemma. *If f is an entire function, then*

$$N(r, f, a) - N(r_0, f, a) \leq T(r, f) + \mathcal{O}(1),$$

where $\mathcal{O}(1)$ is uniform in $a \in \mathbb{P}$.

Proof. Without loss of generality we may assume that $f(0) = 0$. For $a \neq 0$,

$$N(r, f, a) - N(r_0, f, a) \leq N(r, f, a)$$
$$= T(r, f) - m(r, f, a) \leq T(r, f) + \log \frac{1}{\chi(f(0), a)}. \tag{3.5}$$

On the other hand, for a near 0 (recall that $T \geq 0$)

$$N(r, f, a) - N(r_0, f, a) \leq T(r, f) - \big(m(r, f, a) - m(r_0, f, a)\big)$$
$$\leq T(r, f) + \frac{1}{2\pi} \int_0^{2\pi} \log \frac{1}{\chi(f(r_0 e^{i\theta}), a)} d\theta$$
$$\leq T(r, f) + C - \frac{1}{2\pi} \int_0^{2\pi} \log |f(r_0 e^{i\theta}) - a| d\theta \leq T(r, f) + C_1. \tag{3.6}$$

To see the last inequality, note that the integral is $\geq -C_2$ for a near 0 since $f \neq 0$ on $|z| = r_0$ if r_0 is small and the integral is an increasing function of r_0. From (3.5) and (3.6) the lemma follows. $\qquad\square$

Proof of Theorem 3.3. Let $\rho(a)$ be a positive function on $\mathbb{P}$ with total mass 1 with respect to $d\mu(a)$, and let

$$\lambda_\rho(r) = \frac{1}{\pi} \int_0^{2\pi} \rho(f(re^{i\theta})) \frac{|f'(re^{i\theta})|^2}{(1 + |f(re^{i\theta})|^2)^2} d\theta.$$

A principal part of the proof is contained in the following estimate.

3.7 Lemma. *For $r > 0$ outside a set $E \subset [0, \infty)$ of finite measure,*

$$\log \lambda_\rho(r) \leq \log r + 4 \log T(r, f) + \mathcal{O}(1).$$

Proof. First notice that

$$\int_\mathbb{P} n(r, f, a) \rho(a) d\mu(a) = \frac{1}{\pi} \int \frac{\rho(a) d\lambda(a)}{(1 + |a|^2)^2} = \int_0^r \lambda_\rho(t) t \, dt,$$

where the integral in the middle is over the multivalued image of $D(0, r)$ under the mapping f. Hence, in view of Lemma 3.6 we get

$$\int_{r_0}^r \frac{ds}{s} \int_0^s \lambda_\rho(t) t \, dt \tag{3.7}$$
$$= \int_\mathbb{P} (N(r, f, a) - N(r_0, f, a)) \, \rho(a) d\mu(a) \leq T(r, f) + C.$$

To conclude the lemma from (3.7), let

$$L(s) = \int_0^s \lambda_\rho(t) t \, dt \quad \text{and} \quad K(r) = \int_{r_0}^r L(s) \frac{ds}{s}.$$

Let E' be the set where $\lambda_\rho(s) \geq (L(s))^2/s$. If r_1 is such that $L(r_1) > 0$, then

$$\int_{E' \cap [r_1, \infty)} ds \leq \int_{r_1}^\infty \frac{s \lambda_\rho(s) ds}{L(s)^2} = \int_{r_1}^\infty \frac{dL(s)}{L(s)^2} \leq \frac{1}{L(r_1)} < \infty.$$

Similarly, if E'' is the set where $L(r) \geq r K(r)^2$, then

$$\int_{E'' \cap [r_1, \infty)} dr = \int_{E'' \cap [r_1, \infty)} r \frac{dK(r)}{L(r)} \leq \int_{r_1}^\infty \frac{dK(r)}{K(r)^2} \leq \frac{1}{K(r_1)} < \infty.$$

For r outside $E = E' \cup E''$, we have

$$\lambda_\rho(r) \leq \frac{L(r)^2}{r} \leq \frac{r^2 K(r)^4}{r} \leq r(T(r, f) + C)^4,$$

and hence the lemma is proved. $\square$

We now can conclude the proof of Theorem 3.3. By Jensen's inequality,

$$\frac{1}{2\pi} \int_0^{2\pi} \log \left(\rho(f(re^{i\theta})) \frac{|f'(re^{i\theta})|^2}{(1 + |f(re^{i\theta})|^2)^2} \right) d\theta \leq \log \lambda_\rho(r, f) - \log 2,$$

and hence

$$\frac{1}{2\pi} \int_0^{2\pi} \log \rho(f(re^{i\theta}))d\theta \le 4T(r,f) + \log \lambda_\rho(r,f) + \mathcal{O}(1) \qquad (3.8)$$

when $r \to \infty$ since $\log |f'(z)|^2$ is subharmonic. Let $a_0, a_1, \ldots, a_k$ be distinct points on $\mathbb{P}$ and set

$$\log \rho(a) = 2 \sum \log \frac{1}{\chi(a,a_j)} - 2\log \left(\sum \log \frac{2}{\chi(a,a_j)} \right) - C.$$

Then $\rho(a)$ is integrable with respect to $d\mu(a)$, and we can choose C so that the total mass of ρ is 1. We may assume that $f(0)$ differs from each of $a_0, \ldots, a_k$. If we apply (3.8) to this choice of ρ, we get

$$2 \sum m(r,f,a_j) \le 4T(r,f) + \log \lambda_\rho(r)$$
$$+ \frac{1}{\pi} \int_0^{2\pi} \log \left(\sum \log \frac{1}{\chi(f(re^{i\theta}),a_j)} \right) d\theta + \mathcal{O}(1).$$

Jensen's inequality and (3.4) (recall that $N \ge 0$) yield that

$$\frac{1}{2\pi} \int_0^{2\pi} \log \left(\sum \log \frac{1}{\chi(f(re^{i\theta}),a_j)} \right) d\theta \le C \log T(r,f) + \mathcal{O}(1),$$

and hence by Lemma 3.7 we get

$$\sum_0^k m(r,f,a_j) \le 2T(r,f) + \frac{1}{2}\log r + C \log T(r,f) + \mathcal{O}(1) \qquad (3.9)$$

for r outside E. Since we already know that the theorem holds for polynomials, in view of Lemma 3.1 we may assume that

$$\limsup_{r\to\infty} \frac{\log r}{T(r,f)} = 0.$$

Letting $a_0 = \infty$ we now obtain Theorem 3.3 from (3.9) since

$$\delta(f,a) = 1 - \limsup_{r\to\infty} \frac{N(r,f,a)}{T(r,f)} = \liminf_{r\to\infty} \frac{m(r,f,a)}{T(r,f)}$$

(if $f(0) \ne a$) and $\delta(f,a_0) = 1$.   □

Supplementary Exercises

Exercise 1. Suppose that $f \in A(U)$. One says that $\alpha \in T$ is a regular point for f if there are $r > 0$ and $g \in A(D(\alpha,r))$ such that $f = g$ on $U \cap D(\alpha,r)$. Show that if $f = \sum a_n z^n$ and $\limsup |a_n|^{1/n} = 1$, then some point on T is singular (i.e., nonregular).

Exercise 2. Let a_j be points in the upper half-plane. Find a condition on $\{a_j\}$ so that $\prod(z-a_j)/(z-\bar{a}_j)$ defines an analytic function there.

Exercise 3. Prove Jensen's formula (2.2) without any reference to Jensen's formula in Ch. 4. First show that the mean value property for the function

$$\log \left| f(z) \prod \frac{r}{z - \alpha_n} \right|$$

implies the desired equality modulo the sum

$$\sum \frac{1}{2\pi} \int_0^{2\pi} \log |1 - e^{-i\theta} \alpha_n / r| d\theta.$$

Then show that each of these integrals actually vanishes.

Exercise 4. Suppose that α_j is a sequence in $\mathbb{C}$ such that

$$\sum 1/|\alpha_j|^{p+1} < \infty.$$

Show that

$$f(z) = \prod_j W_p(z/\alpha_j)$$

defines an entire function such that $|f(z)| \le \exp C |z|^{p+1}$.

Exercise 5. Show that if f is an entire function of finite order, i.e., $\log |f(z)| = \mathcal{O}(|z|^m)$ for some m when $|z| \to \infty$, then

$$f(z) = z^n e^{g(z)} \prod_j W_p(z/\alpha_j)$$

for some p and n. What is the connection between m and p?

Exercise 6. Suppose that α_j are the zeros of some $f \in A(U)$ and that $|\alpha_1| \le |\alpha_2| \le \cdots \le 1$. Show that $\int_0^1 n(r, f, 0) dr = \sum 1 - |\alpha_j|$.

Exercise 7. Suppose that $f \in H^\infty$ and $f(1 - 1/n) = 0$ for all $n \in \mathbb{Z}^+$. Show that $f \equiv 0$.

Exercise 8. Suppose that $0 < \lambda_1 \le \lambda_2 \le \ldots < \infty$. Show that there is a bounded analytic function $f \not\equiv 0$ in the right half-plane that vanishes at each λ_j if and only if $\sum 1/\lambda_j < \infty$.

Exercise 9. Suppose that $0 < \lambda_1 \le \lambda_2 \le \ldots < \infty$ and that $\sum 1/\lambda_k = \infty$. Show that the space of finite linear combinations of the functions $1, x^{\lambda_1}, x^{\lambda_2}, \ldots$ is dense in $C([0, 1])$. Hint: Use the Hahn–Banach theorem and the preceding exercise.

Exercise 10. Suppose that α_j is a sequence in U. Show that the Blaschke product (2.4) converges (if and) only if α_j satisfies the Blaschke condition.

Exercise 11. Suppose that Ω is finitely connected, h is analytic, and u is a real solution to $\Delta u = \Delta \log |h|$. Take one point b_j from each bounded component of $\mathbb{P} \setminus \Omega$. Show that there are real numbers α_j with $|\alpha_j| \leq \pi$ and a function $f(z)$ with the same zeros as $h(z)$ such that

$$|f(z)| \leq e^{u(z)}\Pi_1^M|z - b_j|^{\alpha_j}.$$

Hint: Determine α_j such that $u - \log|h| - \sum \alpha_j \log|z - b_j|$ has a well-defined harmonic conjugate modulo 2π.

Exercise 12. Show that

$$T(r, f) \leq \frac{1}{2\pi} \int_0^{2\pi} |\log|f(re^{i\theta})||d\theta + \mathcal{O}(1) \leq 2T(r, f) + \mathcal{O}(1)$$

when $r \nearrow R$. Hint: Use the fact that $u = \log|f|$ is subharmonic, so that

$$-\infty < \frac{1}{2\pi} \int_0^{2\pi} u(r_0 e^{i\theta}) \leq \frac{1}{2\pi} \int_0^{2\pi} u^+(re^{i\theta})d\theta - \frac{1}{2\pi} \int_0^{2\pi} u^-(re^{i\theta})d\theta$$

for $r_0 < r < R$.

Exercise 13. Verify the claims in Example 3.2.

Exercise 14. Show that if f is meromorphic in $D(0, R)$ and $f(0) \neq a, b \in \mathbb{P}$, then

$$m(r, f, a) + N(r, f, a) = m(r, f, b) + N(r, f, b), \quad r < R.$$

Exercise 15. Let f be an entire function that attains each value at most m times. Show that f must be a polynomial of degree at most m.

Exercise 16. Define the potential of ρ as

$$p(w) = \int_{\mathbb{P}} \log \frac{1}{\chi(w, a)} \rho(a)d\mu(a),$$

and show that

$$m_\rho(r, f) = \frac{1}{2\pi} \int_0^{2\pi} p(f(re^{i\theta}))d\theta - p(f(0)).$$

Exercise 17. Show that

$$T(r, f) = \int_{\mathbb{P}} N(r, f, a)d\mu(a).$$

Exercise 18. Suppose that f is an entire function that attains two distinct values at most a finite number of times. Show that f must be a polynomial.

Exercise 19. Show that

$$rT'(r, f) = \int_{\mathbb{P}} n(r, f, a)d\mu(a) =: A(r, f).$$

Derive the formula

$$T(r, f) = \int_0^r A(t, f)\frac{dt}{t}.$$

Note that $A(r, f)$ is the area (counted with multiplicities) of the image of the set $\{|z| \le r\}$ under the mapping $z \mapsto f(z)$.

Exercise 20. Show that

$$\frac{1}{2\pi} \int_0^{2\pi} \log^+ |f(re^{i\theta})| d\theta - \log^+ |f(0)| = \frac{1}{2\pi} \int_0^{2\pi} N(r, f, e^{i\theta}) d\theta$$

if f is analytic $D(0, R)$ and $r < R$.

Notes

A thorough treatment of the relationships between zeros, growth, and Taylor coefficients of entire functions can be found in [B].

The value distribution theory was developed by Nevanlinna in the beginning of the 1920s. Proposition 3.5 is called the *first fundamental (or main) theorem*, and formula (3.9) is called the *second fundamental (or main) theorem*. The proof given here is due to Ahlfors. A thorough discussion of these matters, including the more general result for meromorphic functions and further references, can be found in [Hi].

Since we assume that f is analytic and not merely meromorphic, we easily get (3.8), which in turn simplifies the proof of (3.9).

The converse of the result in Exercise 9 is also true; see, e.g., [Ru1] for a proof. This characterization is called the Müntz–Szasz theorem.

6

Harmonic Functions and Fourier Series

§1. Boundary Values of Harmonic Functions

If u is a function in U, we let $u_r(e^{i\theta}) = u(re^{i\theta})$ for $r < 1$. Let

$$\|\phi\|_{L^p} = \left(\frac{1}{2\pi}\int_0^{2\pi}|\phi(e^{it})|^p dt\right)^{1/p}$$

be the L^p-norm on T (for $p < \infty$); and for $1 \le p \le \infty$, let h^p denote the space of harmonic functions in U such that

$$\|u\|_{h^p} = \sup_{r<1}\|u_r\|_{L^p} < \infty.$$

Note that $r \mapsto \|u_r\|_{L^p}$ is increasing since $|u|^p$ is subharmonic, and therefore sup may be replaced by lim. One readily verifies that h^p is a Banach space.

Let $\mathcal{M}(T)$ be the space of complex measures μ on T with norm

$$\|\mu\| = \frac{1}{2\pi}\int_T d|\mu|.$$

It is convenient to identify measures (and functions) on T with 2π periodic measures (and functions) on $\mathbb{R}$, so that, e.g.,

$$\frac{1}{2\pi}\int_T f d\mu = \frac{1}{2\pi}\int_0^{2\pi} f(e^{it})d\mu(t) = \frac{1}{2\pi}\int_0^{2\pi} f(t)d\mu(t),$$

where the integration is to be performed over, e.g., $[0, 2\pi)$. For $\mu \in \mathcal{M}(T)$ we define the Poisson integral of μ as

$$P\mu(re^{i\theta}) = \frac{1}{2\pi}\int_0^{2\pi} P_r(\theta - t)d\mu(t), \quad r < 1;$$

if $d\mu = f dt$ and $f \in C(T)$, this definition is consistent with the previous one in Ch. 4. Note that $P\mu$ is harmonic in U. To begin with, we shall generalize Proposition 1.4 in Ch. 4.

1.1 Theorem.

(a) If $f \in L^p(T)$, $1 \leq p \leq \infty$, then $Pf \in h^p$ and $\|Pf\|_{h^p} = \|f\|_{L^p}$. If $1 \leq p < \infty$, then

$$\lim_{r \nearrow 1} \|(Pf)_r - f\|_{L^p} = 0.$$

If $f \in L^\infty(T)$, then $(Pf)_r \to f$ weak*, i.e.,

$$\int_T (Pf)_r \phi \to \int_T f\phi \quad \text{for all } \phi \in L^1(T).$$

(b) If $\mu \in \mathcal{M}(T)$, then $P\mu \in h^1$, $\|P\mu\|_{h^1} = \|\mu\|$, and $(P\mu)_r \to \mu$ weak*, i.e.,

$$\int (P\mu)_r \phi \to \int \phi d\mu \quad \text{for all } \phi \in C(T).$$

(c) If $f \in C(T)$, then $Pf \in h^\infty \cap C(\overline{U})$, and $(Pf)_r \to f$ uniformly.

Proof. First notice that (c) follows from (the proof of) Proposition 1.4 in Ch. 4. To prove (b) notice that if $\mu \in \mathcal{M}(T)$, then

$$|P\mu(re^{i\theta})| \leq \frac{1}{2\pi} \int_0^{2\pi} P_r(\theta - t)d|\mu|(t);$$

and if we integrate this inequality and use Fubini's theorem, we get that $\|P\mu\|_{h^1} \leq \|\mu\|$. Then take $\phi \in C(T)$. By Fubini's theorem,

$$\frac{1}{2\pi} \int_0^{2\pi} (P\mu)_r(e^{i\theta})\phi(e^{i\theta})d\theta = \frac{1}{2\pi} \int_0^{2\pi} \frac{1}{2\pi} \int_0^{2\pi} P_r(\theta - t)d\mu(t)\phi(e^{i\theta})d\theta$$

$$= \frac{1}{2\pi} \int_0^{2\pi} (P\phi)_r(e^{it})d\mu(t);$$

but since $(P\phi)_r \to \phi$ uniformly according to (c), the right-hand side tends to $\int_0^{2\pi} \phi d\mu/2\pi$. Finally (cf. Exercise 4),

$$\|\mu\| = \sup_{|\phi| \leq 1} \left| \frac{1}{2\pi} \int_0^{2\pi} \phi d\mu \right| = \sup_{|\phi| \leq 1} \lim_r \left| \frac{1}{2\pi} \int_0^{2\pi} \phi(P\mu)_r \right|$$

$$\leq \lim_r \frac{1}{2\pi} \int_0^{2\pi} |(P\mu)_r| = \|P\mu\|_{h^1},$$

and thus (b) is proved. For (a), notice that (Jensen's inequality)

$$\left| \frac{1}{2\pi} \int_0^{2\pi} P_r(\theta - t)f(e^{it})dt \right|^p \leq \frac{1}{2\pi} \int_0^{2\pi} P_r(t) \left| f(e^{i(\theta - t)}) \right|^p dt,$$

for $1 \leq p < \infty$. Integrating this inequality we get $\|Pf\|_{h^p} \leq \|f\|_{L^p}$. Take $\epsilon > 0$ and $\phi \in C(T)$ such that $\|f - \phi\|_{L^p} < \epsilon$. Then

$$\|(Pf)_r - f\|_{L^p} \leq \|(P(f - \phi))_r\|_{L^p} + \|(P\phi)_r - \phi\|_{L^p}$$

$$+ \|\phi - f\|_{L^p} \leq \|f - \phi\|_{L^p} + \|(P\phi)_r - \phi\|_{L^\infty} + \|f - \phi\|_{L^p} < 3\epsilon$$

if r is near 1 because of (c), and hence $(Pf)_r \to f$ in $L^p(T)$; in particular, $\|Pf\|_{h^p} = \lim_r \|(Pf)_r\|_{L^p} = \|f\|_{L^p}$. The case $p = \infty$ is proved analogously to (b). $\qquad\square$

We now consider the converse of Theorem 1.1.

1.2 Theorem. *If $u \in h^p$, $1 < p \le \infty$, there is a unique $f \in L^p(T)$ such that $u = Pf$. If $u \in h^1$, there is a unique $\mu \in \mathcal{M}(T)$ such that $u = P\mu$.*

Thus, we have a one-to-one correspondence (in fact, an isometric isomorphism) between h^p and $L^p(T)$ for $p > 1$ and between h^1 and $\mathcal{M}(T)$. In one direction the correspondence is given by the Poisson integral, $\mu \mapsto P\mu$, and in the other direction by $u \mapsto \lim_r u_r$ in the sense described in Theorem 1.1.

Note that Theorem 1.2 differs from Theorem 1.1 in the sense that it is an existence theorem; it states that an object with certain properties exists. The key is the Banach–Alaoglu theorem, a proof of which is outlined in Exercise 5.

1.3 The Banach–Alaoglu Theorem. *If B is a separable Banach space, $\{\Lambda_\alpha\} \subset B^*$, and $\|\Lambda_\alpha\| \le M$, then there is a subsequence Λ_{α_j} and $\Lambda \in B^*$ such that $\Lambda_{\alpha_j} u \to \Lambda u$ for all $u \in B$, i.e., $\Lambda_{\alpha_j} \to \Lambda$ weak*.*

1.4 Example. In general, $\|\Lambda\| \le \liminf \|\Lambda_j\|$ if $\Lambda_j \to \Lambda$ weak* (Exercise 4) and strict inequality can occur. Take, e.g., $B = L^1(T)$ and let $\Lambda_n = e^{in\theta}$. Then, according to the Riemann–Lebesgue lemma (see next section), $\lim \Lambda_n f = 0$ for all $f \in L^1(T)$, although $\|\Lambda_n\|_{L^\infty} = 1$.

Proof of Theorem 1.2. Suppose that $p > 1$ and $u \in h^p$. Then $\{u_r\}$ is a bounded family of elements in $L^p(T)$ and therefore there is a subsequence u_{r_j} (such that $r_j \nearrow 1$) and $f \in L^p(T)$ such that $\int u_{r_j} \phi \to \int f\phi$ for all $\phi \in L^q(T)$. We have to verify that u is the Poisson integral of f. Note that $z \mapsto u(r_j z)$ is harmonic in a neighborhood of $\overline{U}$ and therefore

$$u(rr_j e^{i\theta}) = \frac{1}{2\pi} \int_0^{2\pi} P_r(\theta - t) u_{r_j}(e^{it}) dt.$$

For fixed $r < 1$ and θ, $t \mapsto P_r(\theta - t)$ is continuous, in particular, in $L^q(T)$, and therefore we get that

$$u(re^{it}) = \frac{1}{2\pi} \int_0^{2\pi} P_r(\theta - t) f(e^{it}) dt,$$

i.e., $u = Pf$. From Theorem 1.1 it now follows that $u_r = (Pf)_r \to f$ and so f is unique; therefore, the proof is complete for $p > 1$. Notice that $L^1(T) \subset \mathcal{M}(T)$, which is the dual of $C(T)$; therefore, if $u \in h^1$, then the Banach–Alaoglu theorem provides a $\mu \in \mathcal{M}(T)$ such that $\int u_{r_j} \phi \to \int \phi d\mu$

for all $\phi \in C(T)$. Then the proof is concluded analogously to the case $p > 1$. $\qquad\square$

If we combine Theorem 1.1 with the observations about the Green potential $\mathcal{G}\mu$ in Remark 2.14 in Ch. 4, we get the following decomposition of a subharmonic function u. The special case when $u = \log|f|$ for some analytic function f will be fundamental in subsequent chapters.

1.5 Theorem (The Riesz Decomposition). *Let u be a subharmonic function such that*

$$\sup_{r<1} \frac{1}{2\pi} \int_0^{2\pi} u^+(re^{i\theta})d\theta < \infty.$$

Then

$$\int_U (1 - |\zeta|)\Delta u < \infty,$$

and there is a unique measure μ on T such that

$$u(z) = \mathcal{G}(\Delta u)(z) + P\mu(z), \qquad z \in U.$$

Moreover, $u_r \to \mu$ weak* and $u_r \to \mu$ in $L^1(T)$ if μ is absolutely continuous.

Proof. According to Corollary 2.16 and Remark 2.14 about the Green potential, $\mathcal{G}(\Delta u)$ is subharmonic and its Laplacian is equal to Δu. It follows (see Exercise 31) that $u - \mathcal{G}(\Delta u)$ is harmonic in U. In view of Exercise 22 in Ch. 4, it is in fact in h^1, and by Theorem 1.2 it is therefore the Poisson integral of a measure μ on T. $\qquad\square$

We now are going to study pointwise convergence of the Poisson integral. To this end we first recall the Lebesgue–Radon–Nikodym decomposition of a measure into an absolutely continuous and a singular part,

$$\mu = \mu_s + \mu_a,$$

i.e., where μ_s is concentrated on a null set (with respect to $d\theta$) and $\mu_a = f\,d\theta$ for some $f \in L^1(T)$. If

$$L_t(s) = \frac{1}{2t}\chi_{-t}^t(s) = \begin{cases} 1/2t, & |s| < t \\ 0, & |s| > t, \end{cases}$$

then

$$\int L_t(\theta - s)d\mu(s) = \frac{1}{2t}\int_{\theta-t}^{\theta+t} d\mu(s) \to f(e^{i\theta}) \text{ for a.e. } \theta \text{ when } t \searrow 0, \quad (1.1)$$

i.e., the mean values over small intervals around $e^{i\theta}$ tend to $f(e^{i\theta})$ for a.e. θ when the lengths tend to zero. If $d\mu = f\,d\theta$ and $f \in C(T)$, this is trivial and can be rephrased as $L_t(s) \to \delta$ weak* as measures when $t \searrow 0$. We

now will consider instead the mean values $(1/2\pi)\int_0^{2\pi} P_r(s - \theta)d\mu(s)$ of μ
when $r \nearrow 1$. Also, $P_r(t)/2\pi \to \delta$ weak* as measures when $r \nearrow 1$. One
can think of $P_r(t)$ as regularized characteristic functions. Figure 1 below
shows $P_r(t)$ for $r = 1/2$, $r = 3/4$, and $r = 7/8$.

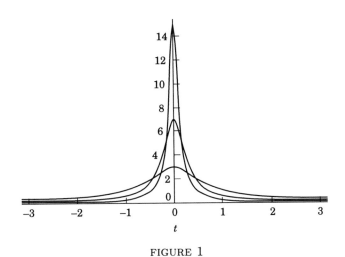

FIGURE 1

One may guess that the analogue of (1.1) holds for $P_r/2\pi$.

1.6 Theorem. *If $\mu = \mu_s + f d\theta$, then*

$$(P\mu)_r \to f \text{ for a.e. } \theta \text{ when } r \nearrow 1. \tag{1.2}$$

In particular,

$$(Pf)_r \to f \text{ for a.e. } \theta \text{ when } r \nearrow 1 \text{ if } f \in L^p(T), \quad 1 \le p \le \infty.$$

Thus, any $u \in h^1$ has radial boundary values u^* a.e.; and if $u = P\mu$,
then u^* is equal to the absolutely continuous part of μ.

Proof. We first recall that the standard proof of (1.1) depends on only
two facts: that (1.1) holds for continuous f and that one has the weak type
(1,1)-estimate,

$$|\{M\mu > \alpha\}| \le \frac{C}{\alpha}\|\mu\|, \tag{1.3}$$

of the Hardy–Littlewood maximal function

$$M\mu(\theta) = \sup_{t>0} \int L_t(\theta - s)d|\mu|(s).$$

Since we know (1.2) for continuous functions, we just have to show the analogue (1.3)' of (1.3), where $M\mu$ is replaced by the new maximal function

$$\tilde{M}\mu(\theta) = \sup_{r<1} \frac{1}{2\pi} \int_0^{2\pi} P_r(\theta - s)d|\mu|(s).$$

We shall reduce (1.3)' to (1.3), and we begin with representing $P_r(s)/2\pi$ as a superposition of the functions $L_t(s)$,

$$\frac{1}{2\pi} P_r(s) = P_r(\pi)L_\pi(s) + \frac{1}{2\pi} \int_0^\pi -P_r'(t)2tL_t(s)dt. \qquad (1.4)$$

In fact, $\int_0^\pi -P_r'(t)2tL_t(s)dt/2\pi = \int_{|s|}^\pi -P_r'(t)dt/2\pi$; an integration by parts then gives (1.4). By (1.4) we get

$\tilde{M}\mu(\theta) =$

$$\sup_{r<1} \left[P_r(\pi) \int L_\pi(s-\theta)d|\mu|(s) + \frac{1}{2\pi} \int_0^\pi -P_t'(t)2t \int L_t(s-\theta)d|\mu|(s)dt \right]$$

$$\leq \sup_{r<1} M\mu(\theta) \left[P_r(\pi) + \frac{1}{2\pi} \int_0^\pi -P_r'(t)2tdt \right] = M\mu(\theta),$$

from which (1.3)' follows, and hence also Theorem 1.6.   □

Recall that for each real harmonic u in U there is a unique analytic function Gu such that $u = \operatorname{Re} Gu$ and Gu is real at the origin.

1.7 Theorem. *Suppose that $1 < p < \infty$. If $u \in h^p$ is real, then its harmonic conjugate $v = \operatorname{Im} Gu$ (which vanishes at the origin) also belongs to h^p and $\|v\|_{h^p} \leq A_p\|u\|_{h^p}$, where A_p depends only on p.*

A conformal mapping of U onto the strip $\{|\operatorname{Re} w| < 1\}$ shows that the theorem is not true for $p = \infty$ and hence (Exercise 8) not true for $p = 1$ either.

A finite sum $\sum a_n e^{in\theta}$ is called a *trigonometric polynomial.* Any real trigonometric polynomial is the restriction to T of the real part of an analytic polynomial $g(z) = \alpha(z) + i\beta(z)$ with $\beta(0) = 0$. The theorem implies in particular that $\|\beta\|_{L^p} \leq A_p\|\alpha\|_{L^p}$.

If $1 < p < \infty$, then

$$Gu(z) = \frac{1}{2\pi} \int_0^{2\pi} \frac{e^{it} + z}{e^{it} - z} u^*(e^{it})dt, \quad |z| < 1,$$

since the integral is analytic, real at the origin, and its real part equals $u = Pu^*$. If we try to estimate the boundary values of Gu by letting $z = e^{i\theta}$ formally, we get $e^{it} - e^{i\theta} = \mathcal{O}(|t - \theta|)$ in the denominator, and therefore that integral does not exist in the usual sense; it is a so-called singular integral. There is a theory for such integrals which can be used to deduce Theorem 1.7, but here we supply a more elementary proof.

Proof. First suppose that $1 < p \leq 2$, $f = u + iv$ is analytic, $v(0) = 0$, and $u > 0$. Since $2\partial u/\partial z = f'$, a simple computation gives that

$$\Delta u^p = p(p-1)|f'|^2 u^{p-2}$$

and

$$\Delta |f|^p = p^2 |f'|^2 |f|^{p-2};$$

therefore,

$$\Delta |f|^p \leq \frac{p}{p-1} \Delta u^p \tag{1.5}$$

(keep in mind that $1 < p \leq 2$). Notice that Green's formula (see A in the preliminaries) gives that

$$r \int_0^{2\pi} \frac{\partial \phi}{\partial r} d\theta = \int_{|\zeta| < r} \Delta \phi d\lambda.$$

From (1.5) we then get

$$\frac{d}{dr} \int_0^{2\pi} |f(re^{i\theta})|^p d\theta \leq \frac{p}{p-1} \frac{d}{dr} \int_0^{2\pi} |u(re^{i\theta})|^p d\theta;$$

but $v(0) = 0$, and thus

$$\int_0^{2\pi} |f(re^{i\theta})|^p d\theta \leq \frac{p}{p-1} \int_0^{2\pi} |u(re^{i\theta})|^p d\theta$$

for $r = 0$ and hence for any $r < 1$. Thus $\|Gu\|_{h^p} \leq C_p \|u\|_{h^p}$ if $0 \leq u \in h^p$, since (by the maximum principle) a nonnegative harmonic function is strictly positive unless it is identically 0. If $u \in h^p$, then u^* is in $L^p(T)$ and hence so are u^{*+} and u^{*-}. Now $u = Pu^{*+} - Pu^{*-}$, where both terms are nonnegative and in h^p. Moreover, $Gu = GPu^{*+} - GPu^{*-}$, and therefore we get that

$$\|Gu\|_{h^p} \leq C_p \|Pu^{*+}\|_{h^p} + C_p \|Pu^{*-}\|_{h^p} \leq 2C_p \|u^*\|_{L^p} = 2C_p \|u\|_{h^p}.$$

This proves the theorem for $1 < p \leq 2$ with $A_p = 2C_p = 2p/(p-1)$.

Now suppose that $2 \leq p < \infty$ and let $p^{-1} + q^{-1} = 1$. Take an arbitrary real trigonometric polynomial $\alpha = \text{Re}\, g$, where $\text{Im}\, g = \beta$ and $\beta(0) = 0$. For such a g, $\beta u_r + \alpha v_r = \text{Im}\, g f_r$ is harmonic and vanishes at the origin, and therefore $\int u(re^{i\theta})\beta(e^{i\theta})d\theta = -\int v(re^{i\theta})\alpha(e^{i\theta})d\theta$. Thus, Hölder's inequality gives

$$\left| \frac{1}{2\pi} \int_0^{2\pi} v_r \alpha d\theta \right| = \left| \frac{1}{2\pi} \int_0^{2\pi} u_r \beta d\theta \right| \leq \|u_r\|_{L^p} \|\beta\|_{L^q} \leq A_q \|u_r\|_{L^p} \|\alpha\|_{L^q},$$

where we have applied the theorem to α and β. The (real) trigonometric polynomials α are dense in (real) $L^q(T)$ (see below), and thus it follows that $\|v_r\|_{L^p} \leq A_q \|u_r\|_{L^p}$ and so $\|v\|_{h^p} \leq A_q \|u\|_{h^p}$. $\qquad \square$

By virtue of Theorems 1.1, 1.2, and 1.7 we can define a bounded operator, the *Hilbert transform*,

$$H: L^p(T) \to L^p(T), \quad 1 < p < \infty,$$

in the following way: For a real ϕ in $L^p(T)$ we let $H\phi$ be the boundary values of the unique harmonic conjugate to $P\phi$ that vanishes at the origin. The operator so defined is certainly real linear, and it then is extended to complex-valued functions by linearity.

§2. Fourier Series

For $\mu \in \mathcal{M}(T)$ we define the *Fourier coefficients*

$$\mathcal{F}\mu(n) = \hat{\mu}(n) = \frac{1}{2\pi} \int_0^{2\pi} e^{-int} d\mu(t), \quad n \in \mathbb{Z}.$$

Note that $|\hat{\mu}(n)| \le \|\mu\|$, and so we have a bounded operator

$$\mathcal{F}: \mathcal{M} \to l^\infty(\mathbb{Z}).$$

Moreover,

$$\mathcal{F}(\bar{\mu})(n) = \overline{\mathcal{F}\mu(-n)}. \tag{2.1}$$

For an integrable function on T,

$$\mathcal{F}f(n) = \hat{f}(n) = \frac{1}{2\pi} \int_0^{2\pi} e^{-int} f(e^{it}) dt;$$

and if f is in $C^1(T)$, then

$$\mathcal{F}\left(\frac{df}{d\theta}\right)(n) = in\mathcal{F}f(n). \tag{2.2}$$

An easy computation reveals that

$$P\mu(re^{i\theta}) = \sum_{-\infty}^{\infty} r^{|n|}\hat{\mu}(n)e^{in\theta} = \sum_1^\infty \hat{\mu}(n)z^n + \hat{\mu}(0) + \sum_1^\infty \bar{z}^n \hat{\mu}(-n). \tag{2.3}$$

Thus, $P\mu$ and hence μ are completely determined by the Fourier coefficients $\hat{\mu}(n)$, and $P\mu$ is analytic in U if and only if $\hat{\mu}(n) = 0$ for all $n < 0$.

To each measure $\mu \in \mathcal{M}(T)$, we associate the trigonometric polynomials

$$\mathcal{F}_N\mu(e^{i\theta}) = \sum_{-N}^{N} \hat{\mu}(n)e^{in\theta}.$$

In view of (2.3), it is natural to ask whether $\mathcal{F}_N\mu$ tends to μ in any reasonable sense when $N \to \infty$. It follows from (2.2) that $\mathcal{F}f(n) = \mathcal{O}(1/|n|^2)$ when $|n| \to \infty$ if $f \in C^2(T)$, and hence the series in (2.3) converges uniformly even for $r = 1$, i.e., $\mathcal{F}_N f \to f$ uniformly. In fact, it is enough that

$f \in C^1(T)$; see Exercise 13. Since $C^2(T)$ is dense in $L^p(T)$ (use, e.g., Theorem 1.1), it follows that the space of trigonometric polynomials is dense in $L^p(T)$. In particular, this implies that $\{e^{in\theta}\}$ is a complete ON-system in $L^2(T)$, and therefore we have Parseval's equality

$$(f, g) = \sum_{-\infty}^{\infty} \hat{f}(n)\overline{\hat{g}(n)}, \tag{2.4}$$

where

$$(f, g) = \frac{1}{2\pi} \int_0^{2\pi} f(e^{it})\overline{g(e^{it})}dt, \tag{2.5}$$

and hence the isometric isomorphism

$$\mathcal{F}: L^2(T) \simeq l^2(\mathbb{Z}).$$

Moreover, $\|\mathcal{F}_N f - f\|_{L^2} \to 0$ when $N \to \infty$. It is also true that $\|\mathcal{F}_N f - f\|_{L^p} \to 0$ for $f \in L^p(T)$ if $1 < p < \infty$; see Exercise 21. In Exercise 40 in Ch. 1 we saw that

$$\mathcal{F}: L^p(T) \to \ell^q(\mathbb{Z}), \quad 1 < p < 2$$

(the Hausdorff–Young inequality). Furthermore, we have the Riemann–Lebesgue lemma, which states that

$$\mathcal{F}: L^1(T) \to c_0(\mathbb{Z}),$$

where $c_0(\mathbb{Z})$ is the space of sequences c_n such that $c_n \to 0$ when $|n| \to \infty$. This follows quite immediately from the fact that the trigonometric polynomials are dense in $L^1(T)$. It is easily verified that

$$\frac{1}{2\pi} \int_0^{2\pi} (\mathcal{F}_N \mu)\bar{\phi}d\theta = \sum_{-N}^{N} \hat{\mu}(n)\overline{\hat{\phi}(n)} = \frac{1}{2\pi} \int_0^{2\pi} \overline{\mathcal{F}_N \phi}d\mu(\theta) \tag{2.6}$$

for $\mu \in \mathcal{M}(T)$, from which it follows that $\mathcal{F}_N \mu \to \mu$ at least in the weak sense that

$$\int_0^{2\pi} \phi \mathcal{F}_N \mu \to \int_0^{2\pi} \phi d\mu, \quad \phi \in C^2(T).$$

If $\phi \in C^\infty(T)$, then, by (2.2), $\mathcal{F}\phi(n) = \mathcal{O}(|n|^{-M})$ for all M when $|n| \to \infty$, and therefore (2.3) converges uniformly even after an arbitrary number of differentiations. Therefore, $P\phi \in C^\infty(\overline{U})$, and in particular $\mathcal{F}_N \phi \to \phi$ in $C^\infty(T)$ (i.e., uniformly with all its derivatives).

2.1 Remark. Let $\mathcal{P}(\mathbb{Z})$ be the space of sequences c_n that are $\mathcal{O}(|n|^M)$ for some M when $|n| \to \infty$. For any distribution $f \in \mathcal{D}'(T)$, let $\mathcal{F}f(n) = f(e^{-in\cdot})$. As T is compact, f has finite order, so $\mathcal{F}f(n) = \mathcal{O}(|n|^M)$ for some M. Therefore,

$$\mathcal{F}: \mathcal{D}'(T) \to \mathcal{P}(\mathbb{Z}). \tag{2.7}$$

Since (2.6) extends to distributions f, it follows that $\mathcal{F}_N f \to f$ in $\mathcal{D}'(T)$. Conversely, one easily sees that if c_n is any sequence in $\mathcal{P}(\mathbb{Z})$, then the corresponding Fourier series converges in $\mathcal{D}'(T)$, and therefore (2.7) is actually an isomorphism.

Let us express the Hilbert transform in terms of the Fourier coefficients. Let u be a real function in $L^2(T)$, and let $v = Hu$. As the harmonic extension of v vanishes at the origin, $\hat{v}(0) = 0$. Furthermore, as (the harmonic extension of) $u + iv$ is analytic, $\hat{u}(n) + i\hat{v}(n) = 0$ for $n < 0$. In view of (2.1), we then must have that

$$\widehat{Hu}(n) = \begin{cases} -i\hat{u}(n), & n > 0 \\ 0, & n = 0 \\ i\hat{u}(n), & n < 0. \end{cases} \tag{2.8}$$

Notice that (2.8) and Parseval's equality (2.4) imply a very simple proof of Theorem 1.7 for $p = 2$ and that the best constant A_2 is 1. From (2.8) and (2.4) it also follows that

$$HHf = -f + \hat{f}(0) \quad \text{and} \quad H^* = -H, \tag{2.9}$$

i.e., $(Hf, g) = -(f, Hg)$ for all $f, g \in L^2(T)$.

We now introduce the spaces $H^p = h^p \cap A(U)$, $1 \leq p \leq \infty$, which are closed (why?) subspaces of h^p, and we first consider the case $p > 1$.

2.2 Theorem. *Suppose that $1 < p \leq \infty$. If $f \in H^p$, then*

$$f^*(e^{i\theta}) = \lim_{r \nearrow 1} f(re^{i\theta})$$

exists for a.e. θ, $f^* \in L^p(T)$,

$$f(z) = (Pf^*)(z) = \frac{1}{2\pi i} \int_T \frac{f^*(\zeta)d\zeta}{\zeta - z}, \tag{2.10}$$

$\|f\|_{H^p} = \|f^*\|_{L^p}$ *and for $p < \infty$ also* $\|f_r - f^*\|_{L^p} \to 0$ *when $r \nearrow 1$. A function in $L^p(T)$ is f^* for some $f \in H^p$ if and only if its negative Fourier coefficients vanish.*

Proof. Everything but (2.10) is already proved. If $r < 1$, then

$$f(rz) = \frac{1}{2\pi i} \int_T \frac{f(r\zeta)d\zeta}{\zeta - z};$$

and since $f_r \to f^*$ in $L^1(T)$, (2.10) follows. $\qquad\square$

Thus, for $p > 1$, H^p may be identified with the closed subspace $\{u \in L^p(T); \hat{u}(n) = 0, n < 0\}$ of $L^p(T)$. In particular,

$$H^2 \simeq l^2(\mathbb{N}) = \{a \in l^2(\mathbb{Z}); a_n = 0 \text{ for } n < 0\},$$

and therefore we have the orthogonal projection

$$S: L^2(T) \to H^2,$$

referred to as the *Szegö projection*. We claim that

$$\widehat{Sf}(n) = \begin{cases} \hat{f}(n), & n \geq 0 \\ 0, & n < 0. \end{cases} \tag{2.11}$$

In fact, since Sf is analytic (in the sequel we identify Sf with its Poisson integral) for any f and $f - Sf$ is orthogonal to the analytic functions with respect to the inner product (2.5), (2.11) follows from Parseval's equality. Now

$$Sf(z) = \sum_0^\infty \hat{f}(n)z^n = \sum_0^\infty \frac{1}{2\pi} \int_0^{2\pi} f(e^{it})e^{-int}z^n \, dt$$

$$= \frac{1}{2\pi} \int_0^{2\pi} \frac{f(e^{it})dt}{1 - e^{-it}z} = \frac{1}{2\pi i} \int_T \frac{f(\zeta)d\zeta}{\zeta - z},$$

and therefore the (harmonic extension of) the Szegö projection is given by the Cauchy integral. However, the Cauchy integral makes sense for any $f \in L^p(T)$, $p \geq 1$.

2.3 Theorem. *Suppose that $1 < p < \infty$. Then*
(a) the Szegö projection extends to a bounded mapping

$$S: L^p(T) \to H^p.$$

(b) any bounded functional on H^p is represented by a unique element h in H^q via the pairing (2.5), and the functional norm is equivalent to the H^q norm of h.

Part (b) can be rephrased as $(H^p)^* = H^q$.

Proof. From (2.8) and (2.11) we get the relation

$$Su = \frac{1}{2}(u + iHu + Pu(0)), \quad u \in L^2(T). \tag{2.12}$$

Since H is bounded on $L^p(T)$, so is S, and hence the first statement is confirmed.

Any bounded functional Λ on H^p extends by the Hahn–Banach theorem to a functional on $L^p(T)$ with the same norm, and it therefore is represented by a $\phi \in L^q(T)$ with $\|\phi\|_{L^q} = \|\Lambda\|$. In particular, $\Lambda f = (f, \phi)$ for $f \in H^p$, but then also $\Lambda f = (f, S\phi)$ for all $f \in H^p$, since $Sf = f$ and $S^* = S$. Thus, $g = S\phi \in H^q$ represents Λ. Then clearly $\|\Lambda\| \leq \|g\|_{H^q}$, and by the boundedness of S, $\|g\|_{H^q} \leq C\|\phi\|_{L^q} = C\|\Lambda\|$ for some constant C that only depends on q. $\square$

Let us now take a look at the space H^1. From Theorems 1.2 and 1.1 we know that if $f \in H^1 \subset h^1$, then $f^*(e^{i\theta})$ exists for a.e. θ and $f = P\mu$, where $\mu = \mu_s + f^* d\theta$. Since f is analytic, we also know that $\hat{\mu}(n) = 0$ for $n < 0$. The crucial fact is that then μ is in fact absolutely continuous and thus equal to $f^* d\theta$.

2.4 The F. & M. Riesz Theorem. *If $\mu \in \mathcal{M}(T)$ and $\hat{\mu}(n) = 0$ for $n < 0$, then μ is absolutely continuous, i.e., $\mu = g d\theta$ for some $g \in L^1(T)$.*

By an obvious argument we then have the following extension of Theorem 2.2.

2.5 Theorem. *If $f \in H^1$, then $f^*(e^{i\theta}) = \lim_{r \nearrow 1} f(re^{i\theta})$ exists for a.e. θ, $f^* \in L^1(T)$,*

$$f(z) = (Pf^*)(z) = \frac{1}{2\pi i} \int_T \frac{f^*(\zeta) d\zeta}{\zeta - z},$$

$\|f\|_{H^1} = \|f^*\|_{L^1}$ and $\|f_r - f^*\|_{L^1} \to 0$ when $r \nearrow 1$. A function in $L^1(T)$ is f^* for some $f \in H^1$ if and only if its negative Fourier coefficients vanish.

On the other hand, the F. & M. Riesz theorem follows immediately from Theorem 2.5. To prove the latter, we must use that f is analytic and not merely harmonic. In the next paragraph we shall see that any $f \in H^1$ can be factorized as $f = gh$, where $f, g \in H^2$; if f is nonvanishing, this is trivial, we just take $g = h = \phi$, where $\phi^2 = f$. Now Theorem 2.5 easily follows:

$$\|f_r - f^*\|_{L^1} = \|g_r h_r - g^* h^*\|_{L^1} \leq \|g_r h_r - g^* h_r\|_{L^1} + \|g^* h_r - g^* h^*\|_{L^1}$$
$$\leq \|g_r - g^*\|_{L^2} \|h_r\|_{L^2} + \|g^*\|_{L^2} \|h_r - h^*\|_{L^2} \to 0 \text{ when } r \nearrow 0,$$

i.e., $f_r \to f^*$ in $L^1(T)$. However, then $f = Pf^*$, and everything follows.

Supplementary Exercises

Exercise 1. Show that h^p is a Banach space, i.e., that $\| \ \|_{h^p}$ is a complete norm on h^p.

Exercise 2. Show that if u is harmonic and positive in U, then $u \in h^1$ and $u = P\mu$, where $\mu \geq 0$.

Exercise 3. Give an example of a $u \in h^p \setminus h^{p'}$ if $1 \leq p < p' \leq \infty$.

Exercise 4. Let Λ_j be a sequence of elements in the dual space B^* to a Banach space B. Show that $\|\Lambda\| \leq \liminf \|\Lambda_j\|$ if $\Lambda_j \to \Lambda$ weak*.

Exercise 5. Fill in the details in the following sketch of a proof of the Banach–Alaoglu theorem. Let $D = \{u_1, u_2, \dots\}$ be an enumeration of a dense subset of B. Take a subsequence $\Lambda_{1,j}$ from $\{\Lambda_\alpha\}$ such that $\lim_{j\to\infty} \Lambda_{1,j} u_1$ exists. Extract from $\{\Lambda_{1,j}\}$ a subsequence $\Lambda_{2,j}$ such that $\lim_{j\to\infty} \Lambda_{2,j} u_2$ exists, and so on. The diagonal sequence $\Lambda_{j,j}$ then converges at each u_k. One then defines Λu_k as this limit and extends by continuity to an element in B^*, which has the desired properties.

Exercise 6. Suppose that $u = P\mu$ and $\|u\|_{h^1} \le \|u^*\|_{L^1}$. Show that μ is absolutely continuous.

Exercise 7. Give an example of a real $u \in h^\infty$ such that its harmonic conjugate is not in h^∞.

Exercise 8. Show that in general the harmonic conjugate v of a real $u \in h^1$ is not in h^1. Is this true if $u = Pu^*$?

Exercise 9. Give an example of a real $u \in h^1$ such that its harmonic conjugate is not in h^1.

Exercise 10. Let $\mu = \mu^+ - \mu^-$ be the Jordan decomposition of the real measure $\mu \in \mathcal{M}(T)$, and let $u = P\mu$. Show that $\int u_r^+ dt \to \int d\mu^+(t)$ when $r \nearrow 1$.

Exercise 11. Suppose that $u = P(u^* + \mu_s)$. Show that $\int u_r^+ \to \int u^{*+}$ if and only if $\mu_s \le 0$.

Exercise 12. Show that the space of trigonometric polynomials is dense in $C(T)$.

Exercise 13. Show that the partial Fourier sums $\mathcal{F}_n f$ tend uniformly to f if $f \in C^1(T)$ (or if $df/d\theta$ is only in $L^2(T)$ in the distribution sense).

Exercise 14. Verify that H^p is a closed subspace of h^p.

Exercise 15. Verify (2.12) by means of the integral representations of $Gu = P(u + iHu)$ and Su.

Exercise 16. Prove the Riemann–Lebesgue lemma, i.e., that $\mathcal{F}$ maps $L^1(T)$ into $c_0(\mathbb{Z})$.

Exercise 17. Show that $\mathcal{F}: \mathcal{M} \to l^\infty(\mathbb{Z})$ is not surjective. Can you find an explicit example of an $a \in l^\infty(\mathbb{Z})$ that is not in the image?

Exercise 18. Let $D_N(t) = \mathcal{F}_N \delta(t) = \sum_{-N}^N e^{int}$. Show that

$$D_N(t) = \frac{\sin((N+1/2)t)}{\sin(t/2)}$$

and that $\|D_N\|_{L^1} \to \infty$ when $N \to \infty$.

Exercise 19. Show that (cf. the preceding exercise)

$$\mathcal{F}_N f(0) = \frac{1}{2\pi} \int_0^{2\pi} f(t) D_N(-t) dt/2\pi$$

and that $f \mapsto \mathcal{F}_N f(0)$ is a bounded linear functional on $C(T)$ with norm equal to $\|D_N\|_{L^1}$. Then show that there are functions $f \in C(T)$ for which $\mathcal{F}_N f(0)$ does not converge when $N \to \infty$. Hint: Use the Banach–Steinhaus theorem and the preceding exercise.

Exercise 20. Show that $\mathcal{F}: L^1(T) \to c_0(\mathbb{Z})$ is not surjective. Hint: Use Exercise 18 and the open mapping theorem for Banach spaces.

Exercise 21. Show that the partial Fourier sums $\mathcal{F}_N f$ tend to f in $L^p(T)$ norm for $f \in L^p(T)$, $1 < p < \infty$. Hint: First show that the set of partial sums is bounded in $L^p(T)$.

Exercise 22. Show that a harmonic function u in U is in h^2 if and only if

$$\int_U (1 - |\zeta|^2)|\nabla u|^2 d\lambda(\zeta) < \infty.$$

Exercise 23. Show that the Szegö projection is unbounded on $L^\infty(T)$ and $L^1(T)$.

Exercise 24. Let $\alpha \in U$. Show that $f \mapsto f^{(m)}(\alpha)$ is a bounded linear functional on H^2 and find the $g \in H^2$ that represents this functional. How large can $|f^{(m)}(\alpha)|$ be if $\|f\|_{H^2} \le 1$?

Exercise 25. Show that if $\mu \in \mathcal{M}(T)$ and $\hat{\mu}(n) = 0$ for all $n \ge N$, then μ is absolutely continuous.

Exercise 26. Let μ be a real measure on T, and show that

$$\limsup_{r \nearrow 1} \frac{1}{2\pi} \int_0^{2\pi} P_r(\theta - s) d\mu(s) \le \limsup_{t \searrow 0} \int L_t(\theta - s) d\mu(s).$$

Show that $P\mu$ has radial limits at each point where the measure derivative of μ exists.

Exercise 27.
(a) Show that if $\mu \ge 0$ on T and

$$\liminf_{t \to 0} \int L_t(\theta - s) d\mu(s) = 0$$

for all θ in an open interval I, then $\mu(I) = 0$.
(b) Suppose that $u \ge 0$ and is harmonic and that $u(re^{i\theta}) \to 0$ when $r \nearrow 1$ for all $e^{i\theta} \ne 1$. Show that $u(re^{i\theta}) = cP_r(\theta)$ for some constant $c \ge 0$.

Exercise 28. Show that $u(z) = \text{Im}[(1 + z)^2/(1 - z)^2]$ has radial limits 0 for all θ. Show that u is not $P\mu$ for any measure μ.

Exercise 29. Show that the Hilbert transform is an isomorphism on $\mathcal{D}'(T)$.

Exercise 30. Is it true that each $f \in A(U)$ is equal to $g^{(m)}$ for some m and some $g \in H^2$?

Exercise 31. Suppose that u and v are subharmonic in Ω and that $\Delta u = \Delta v$. Show that $u - v$ is harmonic in Ω.

Notes

In this chapter as well as in the later ones we restrict ourselves to the unit disk. Most results have analogues in the upper half-plane Π^+. For instance, $h^p(\Pi^+)$ is the space of harmonic functions in Π^+ such that $\sup_{y>0} \int_\mathbb{R} |u(x + iy)|^p dx < \infty$. There is a one-to-one correspondence between $L^p(\mathbb{R})$ and $h^p(\Pi^+)$ for $p > 1$ and between $\mathcal{M}(\mathbb{R})$ and $h^1(\Pi^+)$, which is given by the Poisson integral

$$Pf(x,y) = \frac{1}{\pi} \int_\mathbb{R} \frac{yf(t)dt}{y^2 + (x-t)^2};$$

cf. Exercise 23 in Ch. 3.

A further discussion about weak* compactness and the Banach–Alaoglu theorem can be found in [Ru2].

Theorem 1.6 was first proved by Fatou in 1906 for $p = \infty$. Therefore, results about pointwise convergence at the boundary often are referred to as Fatou theorems. Any $u \in h^1$ actually has *nontangential* boundary values u^*: If $\Gamma(\theta) = \{re^{it} \in U; \ |\theta - t| < 1 - r\}$, then

$$u^*(e^{i\theta}) = \lim_{\Gamma(e^{i\theta}) \ni z \to e^{i\theta}} u(z)$$

exists for a.e. θ; see, e.g., [G]. See also Exercise 28 in the next chapter.

For $1 < p \leq \infty$, there is $C_p > 0$ such that $\|Mf\|_{L^p} \leq C_p\|f\|_{L^p}$, $f \in L^p(T)$, i.e., M is bounded on $L^p(T)$; see, e.g., [G].

Theorem 1.7 was first proved by M. Riesz in 1927. The proof here is due to P. Stein. The Hilbert transform is the simplest example of a singular integral; see [S] for a thorough discussion.

Carleson proved in 1966 that $\mathcal{F}_N f \to f$ for a.e. θ if $f \in L^2(T)$, *Acta Math.*, vol 116 (1966), 135–157. It was generalized to $p > 1$ by Hunt, *Proc. Conf. Edwardsville, Ill. (1967)*, 235–255, Southern Illinios Univ. Press, Carbondale, Ill. (1968).

For further results about Fourier analysis and periodic distributions, see [Hö].

The theorem of F. and M. Riesz is from 1916. For generalizations to $\mathbb{R}^n$, see [S].

7

H^p Spaces

§1. Factorization in H^p Spaces

We extend the definition of the *Hardy spaces* H^p to $p > 0$:

$$H^p = \left\{ f \in A(U); \ \|f\|_{H^p} = \sup_{r<1} \left(\frac{1}{2\pi} \int_0^{2\pi} |f(re^{i\theta})|^p d\theta \right)^{1/p} < \infty \right\}.$$

For $0 < p < 1$, the triangle inequality does not hold; but $\|f + g\|_{H^p}^p \leq \|f\|_{H^p}^p + \|g\|_{H^p}^p$, and therefore H^p is at least a vector space. Moreover,

$$H^\infty \subset H^p \subset H^s \subset \mathcal{N}, \quad \infty > p > s > 0,$$

and therefore the zero set of any $f \in H^p$ satisfies the Blaschke condition. In fact, it is possible to divide out all the zeros without affecting the norm.

1.1 Theorem. *If $f \in \mathcal{N}$ is not identically zero and B is the corresponding Blaschke product, then $f/B \in \mathcal{N}$. If $f \in H^p$, then $f/B \in H^p$ and $\|f/B\|_{H^p} = \|f\|_{H^p}$.*

Proof. Note that $\log^+ |f/B| \leq \log^+ |f| + \log^+ |1/B| = \log^+ |f| - \log |B|$. Since (Theorem 2.3 in Ch. 5) $\log |B_r| \to 0$ in $L^1(T)$ when $r \nearrow 1$, $f/B \in \mathcal{N}$. Now suppose that $f \in H^p$, and let B_n be the first n factors in B. Then clearly $\|f/B_n\|_{H^p} = \|f\|_{H^p}$ since $|B_n(re^{i\theta})| \to 1$ uniformly when $r \nearrow 1$. Moreover, $|f/B_n| \nearrow |f/B|$ when $n \to \infty$, and thus by monotone convergence we have that

$$\frac{1}{2\pi} \int_0^{2\pi} |(f/B)(re^{i\theta})|^p d\theta = \lim_n \frac{1}{2\pi} \int_0^{2\pi} |(f/B_n)(re^{i\theta})|^p d\theta$$

$$\leq \lim_n \|f/B_n\|_{H^p}^p = \|f\|_{H^p}^p,$$

and therefore $\|f/B\|_{H^p} \leq \|f\|_{H^p}$. Since the converse is trivial, the proof is complete. $\qquad\square$

Theorem 1.1 is the key to the H^p-theory.

1.2 Theorem. If $f \in H^p$, $0 < p \le \infty$, then $f^* = \lim_{r \nearrow 1} f_r$ exists for a.e. θ, $\|f\|_{H^p} = \|f^*\|_{L^p}$, and for $p < \infty$ it is also the case that $\|f_r - f^*\|_{L^p} \to 0$ when $r \nearrow 1$.

Proof. Since the case $p \ge 1$ already has been dealt with in Ch. 6, we may assume that $0 < p < 1$. Take $f \in H^p$ and let B be the corresponding Blaschke product. Then f/B is nonvanishing, and hence it is equal to ϕ^m for some $\phi \in A(U)$. From Theorem 1.1 we get that $\|\phi\|_{H^{pm}} = \|f\|_{H^p}^{1/m} < \infty$, and so if $pm > 1$, then $\phi^* = \lim_r \phi_r$ exists for a.e. θ and $\|\phi^*\|_{L^{pm}} = \|\phi\|_{H^{pm}}$ according to Section 1 in Ch. 6. However, $f = B\phi^m$, and therefore f^* exists a.e. and $\|f^*\|_{L^p} = \|B^*(\phi^*)^m\|_{L^p} = \|\phi^*\|_{L^{mp}}^m = \|\phi\|_{H^{mp}}^m = \|f\|_{H^p}$. For $0 < p < 1$, $|f_r|^p + |f^*|^p - |f_r - f^*|^p \ge 0$, and an application of Fatou's lemma yields the last claim. $\qquad\square$

The Nevanlinna condition on f means precisely that $u = \log|f|$ satisfies the hypothesis in Theorem 1.5 in Ch. 6, and hence it admits a Riesz decomposition. Let us repeat the argument anyway. Assume that $f \in \mathcal{N}$ is not identically zero and B is the corresponding Blaschke product. Then $\log|f/B|$ is harmonic and $f/B \in \mathcal{N}$; therefore, by the the mean value property,

$$\sup_{r<1} \frac{1}{2\pi} \int_0^{2\pi} \left|\log|(f/B)(re^{i\theta})|\right| d\theta < \infty,$$

i.e., $\log|f/B|$ is in h^1. Thus, there is a real measure μ on T such that $\log|f/B| = P\mu$. Now

$$f(z) = cB(z)\exp\frac{1}{2\pi}\int_0^{2\pi}\frac{e^{it}+z}{e^{it}-z}d\mu(t), \quad c \in T, \tag{1.1}$$

since both sides are analytic and have the same modulus. Conversely, if μ is a real measure and f is defined by (1.1), then it is in the Nevanlinna class and the representation (1.1) is unique. If we make the Jordan decomposition $\mu = \mu^+ - \mu^-$, we find that $f = g/h$, where $\|g\|_{H^\infty} \le 1$, $\|h\|_{H^\infty} \le 1$, and $h \neq 0$.

1.3 Theorem. If $f \in \mathcal{N}$ and $f \not\equiv 0$, then the radial limits $f^*(e^{i\theta})$ exist for a.e. θ and $\log|f^*| \in L^1(T)$. Moreover,

$$f(z) =$$
$$cB(z)\exp\frac{1}{2\pi}\int_0^{2\pi}\frac{e^{it}+z}{e^{it}-z}\log|f^*(e^{it})|dt\exp\frac{1}{2\pi}\int_0^{2\pi}\frac{e^{it}+z}{e^{it}-z}d\mu(t), \tag{1.2}$$

where μ is a unique real singular measure and $c \in T$.

Proof. Decompose μ in (1.1) as $\mu = \phi d\theta + \mu_s$, where $\phi \in L^1(T)$ and μ_s is singular. Since $|B^*| = 1$ a.e., $\lim\log|f_r| = \phi$ for a.e. θ. If $f \in H^\infty$, then f^*

exists (Theorem 1.2) and consequently $\log|f^*| = \lim \log|f_r| = \phi \in L^1(T)$, and so $f^* \neq 0$ for a.e. θ. However, an arbitrary $f \in \mathcal{N}$ is $f = g/h$ for some $g, h \in H^\infty$, and since $h^* \neq 0$ a.e., $f^* = g^*/h^*$ exists a.e. and $\log|f^*| = \lim \log|f_r|$. $\qquad\square$

1.4 Corollary. *If $f, g \in \mathcal{N}$ and $f^* = g^*$ on a set with positive measure, then $f \equiv g$.*

Proof. If $f \in \mathcal{N}$ and $f^* = 0$ on a set with positive measure, then $\log|f^*| \notin L^1(T)$ and thus $f \equiv 0$. From Exercise 2 it follows that $f - g \in \mathcal{N}$ if $f, g \in \mathcal{N}$, and hence the corollary follows. $\qquad\square$

1.5 Proposition. *If $f \in H^p$, $0 < p \leq \infty$, then $\mu \leq 0$, if μ is defined by (1.2).*

The class of all $f \in \mathcal{N}$ such that $\mu \leq 0$ is denoted $\mathcal{N}^+$. Thus the proposition says that $H^p \subset \mathcal{N}^+$.

Proof. To begin with, we may assume that f is nonvanishing, i.e., $B \equiv 1$. Then $\log|f| = P\mu$, and we have to prove that the singular part of μ is nonpositive. After taking an mth root if necessary, we also may assume that $f \in H^1$. Now, $|\log^+ x - \log^+ y| \leq |x - y|$ (check!), and since $f_r \to f^*$ in $L^1(T)$, $u_r^+ \to u^{*+}$ in $L^1(T)$ if $u = \log|f|$. By the Banach–Alaoglu theorem some subsequence $u_{r_j}^-$ tends weak* to a positive measure λ when $r_j \nearrow 1$. Since $u_{r_j} \to \mu$ (Theorem 1.1 in Ch. 6), we get

$$u^{*+} - \lambda = \mu = \mu^+ - \mu^-.$$

By the minimality of the Jordan decomposition it now follows that $\mu^+ \leq u^{*+}$. Thus, μ^+ is absolutely continuous and the proof is complete. $\qquad\square$

Definition. A function $M \in H^\infty$ is an *inner function* if $|M^*| = 1$ for a.e. θ.

1.6 Proposition. *A function $M \in H^p$ (or in $\mathcal{N}^+$) is an inner function if and only if it has the form*

$$M(z) = cB(z) \exp \frac{1}{2\pi} \int_0^{2\pi} \frac{e^{it} + z}{e^{it} - z} d\mu(t), \qquad (1.3)$$

where μ is singular and ≤ 0.

Proof. If M has the form (1.3), then it is clear that $|M^*| = 1$ and $\|M\|_{H^\infty} = 1$. Conversely, if M is inner and it is decomposed as in (1.2), then (1.3) follows from Proposition 1.5. $\qquad\square$

Definition. We say that an inner function S is *singular* if its Blaschke product is a constant, i.e., if S is nonvanishing.

Hence, any inner function M has a unique factorization $M = BS$, where B is its Blaschke factor and S is a singular inner function.

Definition. If $\phi \geq 0$ and $\log \phi \in L^1(T)$, then

$$Q(z) = \exp \frac{1}{2\pi} \int_0^{2\pi} \frac{e^{it} + z}{e^{it} - z} \log \phi(e^{it}) dt \qquad (1.4)$$

is called an *outer function*.

Notice that $|Q^*| = \phi$ for a.e. θ according to (1.2).

1.7 Proposition. *If Q is an outer function, then $Q \in H^p$ if and only if $\phi \in L^p(T)$, $0 < p \leq \infty$. In this case, $\|Q\|_{H^p} = \|Q^*\|_{L^p} = \|\phi\|_{L^p}$.*

Proof. First assume that $\phi \in L^p(T)$. By Jensen's inequality,

$$|Q(re^{i\theta})|^p = \exp \frac{1}{2\pi} \int_0^{2\pi} P_r(\theta - t) \log \phi^p dt \leq \frac{1}{2\pi} \int_0^{2\pi} P_r(\theta - t) \phi^p dt.$$

Integrating with respect to θ and applying Fubini's theorem, we get that $Q \in H^p$ and $\|Q\|_{H^p} \leq \|\phi\|_{L^p}$. Conversely, if $Q \in H^p$, then $\|Q^*\|_{L^p} = \|Q\|_{H^p}$ by Theorem 1.2; but $\log|Q^*| = \log \phi$, and therefore $\phi \in L^p(T)$ and $\|\phi\|_{L^p} = \|Q^*\|_{L^p}$. $\qquad \square$

We now are prepared for the main theorem.

1.8 Main Theorem. *Any $f \in N^+$ (in particular, $f \in H^p$) has a unique factorization*

$$f = M_f Q_f = B_f S_f Q_f,$$

where $M_f = B_f S_f$ is an inner function, B_f and S_f are its Blaschke factor and singular factor, respectively, and Q_f is the outer function

$$Q_f(z) = \exp \frac{1}{2\pi} \int_0^{2\pi} \frac{e^{it} + z}{e^{it} - z} \log|f^*| dt. \qquad (1.5)$$

Moreover, $|f^*| = |Q^*|$ for a.e. θ and $f \in H^p$ if and only if $Q_f \in H^p$, which holds if and only if $f^* \in L^p(T)$. In that case,

$$\|f\|_{H^p} = \|Q_f\|_{H^p} = \|f^*\|_{L^p}.$$

If $f \in N$, we have $f = M_f Q_f / \tilde{S}_f = B_f S_f Q_f / \tilde{S}_f$, where $\tilde{S}_f$ also is a singular inner function.

Proof. The existence and uniqueness of the factorization and the equality $|f^*| = |Q^*|$ follow from Theorem 1.3. In view of Theorem 1.7, $Q_f \in H^p$ if and only if $f^* \in L^p(T)$. If $f \in H^p$, then $f^* \in L^p(T)$ by Theorem 1.2, and thus $Q_f \in H^p$. Since $|f| \leq |Q_f|$ if $f \in \mathcal{N}^+$, it follows that $f \in H^p$ if $Q_f \in H^p$. The norm equalities now follow from Theorem 1.2. □

If $f \in \mathcal{N}$ and $f^* \in L^p(T)$, it does not follow in general that $f \in H^p$. If $f = 1/S$, where S is a singular inner function, then $|f^*| = 1$ a.e.; but nevertheless, f is not in any H^p-space, not even in $\mathcal{N}^+$, unless it is constant, because of the uniqueness of the factorization.

1.9 Example. Taking μ as minus the point mass at $e^{it} = 1$, we get the singular inner function

$$S(z) = \exp\left(-\frac{1+z}{1-z}\right).$$

One easily sees that $|S^*(e^{it})| = 1$ for $e^{it} \neq 1$ but 0 for $e^{it} = 1$. Thus,

$$f(z) = 1/S(z) = \exp\frac{1+z}{1-z}$$

is a function in $\mathcal{N} \setminus \mathcal{N}^+$ such that $|f^*(e^{it})| = 1$ for all $e^{it} \neq 1$.

§2. Invariant Subspaces of H^2

We now are going to use the main theorem in Section 1 to study certain subspaces of H^2. If ϕ is an inner function, then $\phi H^2 = \{\phi f;\ f \in H^2\}$ is a closed subspace of H^2. In fact, since $\|\phi f\|_{H^p} = \|f\|_{H^p}$, it is clear that $\phi H^2 \subset H^2$. If $\phi f_j \to g$ in H^2, then ϕf_j is a Cauchy sequence in H^2; but $\|f_j - f_k\|_{H^2} = \|\phi f_j - \phi f_k\|_{H^2}$. Hence, f_j is a Cauchy sequence in H^2, and therefore $f_j \to f$ for some $f \in H^2$; therefore, $g = \phi f \in \phi H^2$.

If $(Sf)(z) = zf(z)$, then clearly

$$S: H^2 \to H^2$$

is a bounded operator with norm 1. If ϕ is an inner function, then $S\phi H^2 \subset \phi H^2$, and therefore ϕH^2 is a closed S-invariant subspace of H^2. The converse is also true.

2.1 Theorem. If V is a closed S-invariant subspace of H^2, then $V = \phi H^2$ for some inner function ϕ.

Proof. Take $f \in V$ with minimal order of its zero at the origin, i.e., such that g/f is analytic at the origin for all $g \in V$. Then $f \notin zV = \{zg;\ g \in V\}$, and therefore zV is a proper closed subspace (for the same reasons as above) of V. Thus, there is an element $\phi \in V$ such that $\phi \perp zV$ and $\|\phi\|_{H^2} = 1$. However, since V is S-invariant, zV is also and hence $\phi \perp z^n V$ for all

$n > 0$, implying that

$$\frac{1}{2\pi} \int_0^{2\pi} |\phi^*(e^{it})|^2 e^{-int} dt = 0, \quad n = 1, 2, 3 \ldots .$$

The function $|\phi^*|^2$ is real and in $L^1(T)$, and therefore it follows that it is equal to a constant a.e. Moreover, since $\|\phi^*\|_{L^2} = 1$, $|\phi^*| = 1$ a.e. and hence ϕ is an inner function. As $z^n\phi \in V$ for all $n \geq 0$, $p\phi \in V$ for all polynomials p. However, the polynomials are dense in H^2 (partial sums of power series converge in the H^2-norm; see below), and therefore $\phi H^2 \subset V$. It remains to show that this inclusion is an equality. To this end, it is enough to show that if $h \in V$ and $h \perp \phi H^2$, then $h = 0$. If h is such a function, then $h \perp \phi z^n$ for $n = 0, 1, 2, \ldots$, i.e.,

$$0 = \frac{1}{2\pi} \int_0^{2\pi} h^* \bar{\phi}^* e^{-int} dt, \quad n = 0, 1, 2, \ldots .$$

However, since $h \in V$, $z^n h \in zV$ for $n = 1, 2, \ldots$ and $\phi \perp zV$, we therefore have

$$0 = \frac{1}{2\pi} \int_0^{2\pi} h^* \bar{\phi}^* e^{int} dt, \quad n = 1, 2, \ldots .$$

Hence, all Fourier coefficients of $h^*\bar{\phi}$ vanish and therefore $h^*\bar{\phi}^* = 0$ a.e. However, since $|\phi^*| = 1$ a.e., it follows that $h^* = 0$ a.e. and hence $h = 0$. $\square$

We claim that $\phi H^2 \subset \psi H^2$ if and only if ϕ/ψ is an inner function. In particular, $\phi H^2 = \psi H^2$ if and only if $\phi = c\psi$ for some $c \in T$. In fact, if $\phi H^2 \subset \psi H^2$, then $\phi = \psi M_h Q_h$ for some $h \in H^2$, and thus $\phi = \psi M_h$ because of the uniqueness of the factorization in inner and outer functions. Conversely, if ϕ/ψ is an inner function, then certainly $\phi H^2 \subset \psi H^2$. If $\phi H^2 = \psi H^2$, then both ϕ/ψ and ψ/ϕ are inner functions, and hence ϕ/ψ is constant.

Notice that ϕ/ψ is an inner function if and only if B_ϕ/B_ψ and S_ϕ/S_ψ are analytic, which in turn means that each zero of ψ is also a zero of ϕ (counted with multiplicity) and that $d\mu_\phi - d\mu_\psi$ is a negative singular measure. Thus the family of closed $\mathcal{S}$-invariant subspaces is a partially ordered set with a rather complicated structure.

2.2 Remark. Recall that H^2 is isometrically isomorphic to

$$l^2(\mathbb{N}) = \left\{ \alpha = (\alpha_0, \alpha_1, \ldots); \ \|\alpha\|_{\ell^2}^2 = \sum |\alpha_j|^2 < \infty \right\}$$

(cf. Section 2 of Ch. 6) if $f = \sum_0^\infty \alpha_n z^n$ is identified to $(\alpha_0, \alpha_1, \ldots) \in \ell^2(\mathbb{N})$. The operator $\mathcal{S}$ corresponds to the *shift operator*

$$S(\alpha_0, \alpha_1, \ldots) = (0, \alpha_0, \alpha_1, \ldots)$$

on $l^2(\mathbb{N})$.

We conclude this section with an approximation theorem. If $g \in H^2$, let $\mathcal{P}(g)$ be the closure of $\{pg;\ p \text{ polynomial}\}$, which certainly is a closed S-invariant subspace. We already know that $\mathcal{P}(1) = H^2$.

2.3 Theorem. *Let f and g be in H^2. Then*
(i) $\mathcal{P}(g) = M_g H^2 = \mathcal{P}(M_g)$.
(ii) $f \in \mathcal{P}(g)$ if and only if M_f/M_g is an inner function.
(iii) $\mathcal{P}(f) = \mathcal{P}(g)$ if and only if $M_f = cM_g$.
(iv) $\mathcal{P}(f) = H^2$ if and only if f is an outer function.

Proof. To begin with, $\mathcal{P}(g) \subset M_g H^2$ since $g \in M_g H^2$ and $\mathcal{P}(g)$ is the least S-invariant closed subspace that contains g. By Theorem 2.1, $\mathcal{P}(g) = \phi H^2$ for some inner function ϕ. However, then $M_g Q_g = g = \phi M_h Q_h$; so $M_g = \phi M_h$ and hence $M_g H^2 \subset \phi H^2 = \mathcal{P}(g)$. We therefore obtain the first equality in (i). The second equality follows if g is replaced by M_g. The other statements are immediate consequences. □

§3. Interpolation in H^∞

In Corollary 1.6 in Ch. 5 we saw that for a given sequence of points a_j in U with no limit points, and complex numbers β_j, one can find an $f \in A(U)$ such that $f(a_j) = \beta_j$. One says that a_j is an *interpolation sequence* (for H^∞) if for each $\beta = (\beta_0, \beta_1, \dots) \in l^\infty$ there is an $f \in H^\infty$ that interpolates, i.e., such that $f(a_j) = \beta_j$. It is easily verified that any interpolation sequence must satisfy the Blaschke condition (Exercise 20). However, the Blaschke condition is not sufficient. The points also must be sufficiently spread out in U.

3.1 Theorem (Carleson). *The sequence a_j is an interpolation sequence for H^∞ if and only if there is a δ such that for each fixed k*

$$\prod_{j \neq k} \left| \frac{a_j - a_k}{1 - \bar{a}_j a_k} \right| \geq \delta. \tag{3.1}$$

Proof of the Necessity. Suppose that a_j is an interpolation sequence and let $T: H^\infty \to l^\infty$ be the bounded linear operator defined by $(Tf)_j = f(a_j)$. Now, a_j being an interpolation sequence means precisely that T is surjective. The quotient space $H^\infty/\operatorname{Ker} T$ is a Banach space with the usual quotient norm, and the induced operator $\tilde{T}: H^\infty/\operatorname{Ker} T \to l^\infty$ is thus bounded, injective, and surjective; and so by the open mapping theorem there is $M > 0$ such that

$$\inf_{k \in \operatorname{Ker} T} \|f + k\|_{H^\infty} = \|f + \operatorname{Ker} T\|_{H^\infty} \leq \frac{M}{2} \|\beta\|_{l^\infty}$$

if $Tf = \beta$. Hence, for each $\beta \in l^\infty$ there is an $f \in H^\infty$ such that $Tf = \beta$ and $\|f\|_{H^\infty} \leq M\|\beta\|_{\ell^\infty}$. In particular, there are f_k with $\|f_k\|_{H^\infty} \leq M$ such that $f_k(a_j) = \delta_{jk}$. Let

$$F_k(z) = f_k(z) \Big/ \prod_{j \neq k}^{\infty} \frac{a_j - z}{1 - \bar{a}_j z} \frac{|a_j|}{a_j}.$$

Then $\|F_k\|_{H^\infty} \leq M$ (since we can divide out the zeros without affecting the norm), and if we let $z = a_k$, we get

$$\prod_{j \neq k} \left| \frac{a_j - a_k}{1 - \bar{a}_j a_k} \right| \geq 1/M.$$

$\square$

The hard part of Theorem 3.1 is to prove that the condition (3.1) is sufficient. The proof we give here is elementary and completely constructive. In Exercise 21 in Ch. 8, a proof is outlined where one first takes an appropriate smooth solution to the problem and then modifies it to an analytic solution by solving a $\partial/\partial \bar{z}$-equation.

First, we reformulate condition (3.1). We say that the sequence a_j is *separated* if

$$\left| \frac{a_j - a_k}{1 - \bar{a}_j a_k} \right| \geq c > 0, \quad j \neq k,$$

uniformly.

3.2 Lemma. *The sequence a_j satisfies the condition (3.1) if and only if it is separated and there is $C_\delta > 0$ such that for all k*

$$\sum_{j=1}^{\infty} \frac{(1 - |a_j|^2)(1 - |a_k|^2)}{|1 - \bar{a}_j a_k|^2} \leq C_\delta. \tag{3.2}$$

Proof. If $0 < c \leq x \leq 1$, then $1 - x \leq -\log x \leq C(1 - x)$ for some C. If the terms in the sum in (3.2) are denoted A_{jk}, then

$$1 - A_{jk} = \frac{|a_j - a_k|^2}{|1 - \bar{a}_j a_k|^2}.$$

If $c \leq 1 - A_{jk} < 1$, then

$$\sum_{j \neq k} A_{jk} \leq -\sum_{j \neq k} \log(1 - A_{jk}) \leq C \sum_{j \neq k} A_{jk},$$

and hence the lemma follows. $\square$

Proof of the Sufficiency. It is enough to find $F_k(z)$ such that

$$F_k(a_j) = \delta_{jk} \tag{3.3}$$

and

$$\sum_1^\infty |F_k(z)| \le C, \tag{3.4}$$

since then $f(z) = \sum_1^\infty \beta_k F_k(z)$ solves the problem.

A first attempt to find such F_k would be

$$F_k(z) = \prod_{j \ne k} \frac{z - a_j}{1 - \bar{a}_j z} \Big/ \frac{a_k - a_j}{1 - \bar{a}_j a_k},$$

which converges, since a_j satisfies the Blaschke condition; in fact, $F_k(z) = B_k(z)/B_k(a_k)$ if B_k denotes the Blaschke product with zeros in $\{a_j;\ j \ne k\}$. The problem is that these F_k do not satisfy (3.4). Therefore we shall modify the construction with weight factors and let

$$F_k(z) = \prod_{j \ne k} \left(\frac{z - a_j}{1 - \bar{a}_j z} \Big/ \frac{a_k - a_j}{1 - \bar{a}_j a_k} \right) \left(\frac{1 - |a_k|^2}{1 - \bar{a}_k z} \right)^2 W(a_k, z), \tag{3.5}$$

where $W(\zeta, z)$ is analytic in z and $W(\zeta, \zeta) = 1$. Then F_k still satisfy (3.3), and we are going to show that W can be chosen so that (3.4) also is satisfied.

We assume that $\ldots |a_k| \le |a_{k+1}| \le \ldots$ and first define

$$\psi(\zeta, z) = \sum_{|a_j| \ge |\zeta|} \left(\frac{1 + \bar{a}_j z}{1 - \bar{a}_j z} - \frac{1 + \bar{a}_j \zeta}{1 - \bar{a}_j \zeta} \right) (1 - |a_j|^2), \quad \zeta, z \in U.$$

Since $\sum 1 - |a_j| < \infty$, the series converges uniformly for fixed ζ and $|z| \le r < 1$, and therefore $z \mapsto \psi(\zeta, z)$ is analytic and $\psi(\zeta, \zeta) = 0$. Moreover,

$$\operatorname{Re} \psi(\zeta, z) = \sum_{|a_j| \ge |\zeta|} \left(\frac{1 - |a_j|^2 |z|^2}{|1 - \bar{a}_j z|^2} - \frac{1 - |a_j|^2 |\zeta|^2}{|1 - \bar{a}_j \zeta|^2} \right) (1 - |a_j|^2).$$

Observe that $1 - |a_j|^2 |\zeta|^2 \le 2(1 - |\zeta|^2)$ if $|a_j| \ge |\zeta|$; so if $W = \exp(-\psi)$, then by (3.2),

$$|W(a_k, z)| \le \exp 2C_\delta \exp \left(- \sum_{|a_j| \ge |a_k|} \frac{(1 - |a_j|^2)(1 - |a_j|^2 |z|^2)}{|1 - \bar{a}_j z|^2} \right) \tag{3.6}$$

and $W(a_k, a_k) = 1$.

From (3.1), (3.5), and (3.6) we now get that

$$|F_k(z)| \le \frac{\exp 2C_\delta}{\delta} c_k \exp \left(- \sum_{j \ge k} c_j \right),$$

where

$$c_k = \left(\frac{1 - |a_k|^2}{|1 - \bar{a}_k z|} \right)^2.$$

If $c_1, c_2, \ldots$ are any positive numbers, then

$$c_k \exp\left(-\sum_{j \geq k} c_j\right) \leq \int_{\sum_{j \geq k+1} c_j}^{\sum_{j \geq k} c_j} e^{-t} dt;$$

therefore,

$$\sum_k |F_k(z)| \leq C \sum_k c_k \exp\left(-\sum_{j \geq k} c_j\right) \leq C \int_0^\infty e^{-t} dt = C,$$

and thus (3.4) holds. □

We now are going to study the meaning of the condition (3.1). To begin with, let us assume that all of the points lie on the positive real axis, i.e., $a_k = r_k = 1 - \epsilon_k$ and $0 \leq r_1 < r_2 < \ldots < 1$. Notice that $1 - r_k^2 \sim 1 - r_k$ and $1 - r_k r_j \sim \max(1 - r_j, 1 - r_k)$. If r_k is separated, we thus have that

$$c' \leq \left|\frac{r_k - r_{k+1}}{1 - r_k r_{k+1}}\right| \sim \frac{\epsilon_k - \epsilon_{k+1}}{\epsilon_k},$$

and therefore $\epsilon_{k+1} \leq c\epsilon_k$ for some $c < 1$. Conversely, if this inequality holds, then r_k actually satisfies (3.1). In fact, then it is certainly separated and since $\epsilon_k \leq c^{k-j}\epsilon_j$ if $k \geq j$, we have that

$$\sum_{j=0}^\infty \frac{(1-r_j^2)(1-r_k^2)}{(1-r_j r_k)^2} \sim \sum_{j \leq k} \frac{1-r_k}{1-r_j} + \sum_{j > k} \frac{1-r_j}{1-r_k}$$
$$\lesssim \sum_{j \leq k} c^{k-j} + \sum_{j > k} c^{j-k} \lesssim 1,$$

independently of k. By Lemma 3.2, (3.1) therefore holds. Notice that, for example, $r_k = 1 - 1/k^2$ satisfies the Blaschke condition but is not separated, and hence it is not an interpolation sequence. In the next section we shall see, however, that if r_k satisfies the Blaschke condition, one can always choose θ_k such that the sequence $a_k = r_k e^{i\theta_k}$ satisfies (3.1) and hence is an interpolation sequence.

§4. Carleson Measures

The interpolation condition (3.1) does not only involve the modulii of the points a_k, but also means that they are spread out in U in a certain sense. In order to understand this further one has to introduce the concept of a Carleson measure. If $z, \zeta \in U$, then

$$|1 - \bar\zeta z| \sim \max(1 - |z|, 1 - |\zeta|) + |\arg z - \arg \zeta|, \qquad (4.1)$$

if the arguments are chosen so that the last term is less than or equal to π. This simple observation will be used repeatedly in this section. Let

$$\omega(e^{i\theta}, t) = \{\zeta \in U; \ |1 - e^{-i\theta}z| \leq t\}, \quad t > 0$$

be the so-called *Carleson tents* at $e^{i\theta}$. A positive measure μ in U is a *Carleson measure* if

$$\mu(\omega(e^{i\theta}, t)) \leq Ct,$$

uniformly for all t and θ. The infimum of all possible constants is called the Carleson norm of μ.

4.1 Proposition. *A positive measure μ in U is a Carleson measure if and only if*

$$\int_U \frac{1 - |z|^2}{|1 - \bar{\zeta}z|^2} d\mu(\zeta) \leq C \tag{4.2}$$

uniformly in $z \in U$.

The exact shape of the tents is not important. In this context it is often convenient to use dyadic cubes and a *dyadic decomposition* of U. Let

$$Q_{k,\ell} = \{re^{i\theta}; \ 2^{-k}(\ell - 1) \leq \theta/\pi < 2^{-k}\ell, \ 0 < 1 - r < 2^{-k}\},$$

where $k = 0, 1, 2, \ldots$ and $\ell = -2^k + 1, -2^k + 2, \ldots - 1, 0, 1, \ldots, 2^k$. Then each set $\omega(e^{i\theta}, t)$ is contained in the union of at most two dyadic cubes $Q_{k,\ell}$ with $t \leq C2^{-k}$ for some absolute constant C (cf. (4.1)), and therefore μ is a Carleson measure if and only if $\mu(Q_{k,\ell}) \leq C2^{-k}$ for all k and ℓ. We also introduce the sets

$$B_{k,\ell} = \{re^{i\theta}; \ 2^{-k}(\ell - 1) \leq \theta/\pi < 2^{-k}\ell, \ 2^{-(k+1)} \leq 1 - r < 2^{-k}\},$$

which constitute a partition of U; cf. Figure 1.

Proof of Proposition 4.1. First assume that (4.2) holds and take an arbitrary cube $Q_{k,\ell}$. If $z \in B_{k,\ell}$ and $\zeta \in Q_{k,\ell}$, then $1 - |z|^2 \sim 2^{-k}$ and $|1 - \bar{\zeta}z| \sim 2^{-k}$ (cf. (4.1)), and if we restrict the integration in (4.2) to $Q_{k,\ell}$, we thus get that $\mu(Q_{k,\ell}) \leq C2^{-k}$. Conversely, fix $z \in U$. By invariance we may assume that $z \in B_{k_0,1}$. If $\zeta \in Q_{k_0,\ell}$, then $|1 - \bar{\zeta}z| \sim 2^{-k_0}(1 + |\ell|)$ by (4.1), and hence

$$\sum_\ell \int_{Q_{k_0,\ell}} \frac{1 - |z|^2}{|1 - \bar{\zeta}z|^2} d\mu(\zeta) \sim \sum_\ell \frac{1}{(1 + |\ell|)^2} 2^{k_0} \mu(Q_{k_0,\ell}) \leq C,$$

independently of k_0. It remains to estimate the integral over the boxes $B_{k,\ell}$ for $k < k_0$. However, on each such box we have the trivial estimate $\mu(B_{k,\ell}) \leq \mu(Q_{k,\ell}) \leq C2^{-k}$ and, moreover, $|1 - \bar{\zeta}z| \sim 2^{-k}(1 + |\ell|)$ if $\zeta \in B_{k,\ell}$,

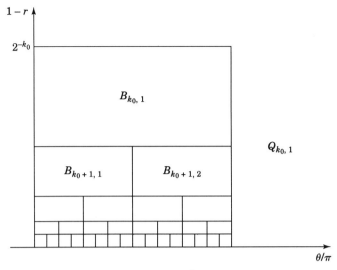

FIGURE 1

and therefore we get

$$\sum_{\ell}\sum_{k<k_0}\frac{2^{-k_0}}{2^{-2k}(1+|\ell|)^2}C2^{-k}\leq C.$$

□

Condition (3.2) means precisely that $\mu = (1 - |\zeta|)\sum\delta_{a_j}$ is a Carleson measure. In one direction it is obvious since condition (4.2) means that (3.2) holds for all z; in particular, for $z = a_k$. For the other direction, fix a dyadic cube. Again we may assume that it is $Q_{k_0,1}$. If there is some $a_k \in B_{k_0,1}$, then by (3.2)

$$\sum_{a_j\in Q_{k_0,1}}\frac{(1-|a_j|^2)(1-|a_k|^2)}{|1-\bar{a}_ja_k|^2}\leq C,$$

which implies that $\mu(Q_{k_0,1}) = \sum_{a_j\in Q_{k_0,1}}(1 - |a_j|) \leq C2^{-k_0}$. If there is no a_k in $B_{k_0,1}$, we instead can apply the same argument to the two cubes below $B_{k_0,1}$ and so on. In this way we obtain an exhausting sequence of subsets of $Q_{k_0,1}$ in which the measure of each set is bounded by $C2^{-k_0}$.

4.2 Example. Now suppose that $0 < r_1 \leq r_2 \leq \ldots < 1$ and $\sum(1 - r_j) < \infty$. We claim that one can choose θ_j such that $a_j = r_je^{i\theta_j}$ is an interpolation sequence. Let $E_k = \{r_i;\ 2^{-(k+1)} \leq 1 - r_j < 2^{-k}\}$, and let c_k be its cardinality. Then $\sum_k 2^{-k}c_k < \infty$. Notice that c points always can be distributed into two boxes so that no box contains more than $c/2 + 1/2$ points. Now fix a k. We can choose points $a_j = r_je^{i\theta_j}$, $r_j \in E_k$, such that the union of all $B_{k,\ell}$ that are contained in the same $Q_{k',\ell'}$ contains at most $2^{-k'}c_k + \sum_{i=1}^{k'} 2^{-i}$ points. In fact, first arrange the points so that at most $c_k/2 + 1/2$ of them lie in each $Q_{0,\ell'}$, $\ell' = 0, 1$; then proceed so that at most $(c_k/2 + 1/2)/2 + 1/2$ points occur in each $Q_{1,\ell'}$, and so on. Let $\tau = \sum \delta_{a_j}$ and $\mu = (1 - |z|)\tau$. If now Q_{k_0,ℓ_0} is a fixed cube, then

$$\mu(Q_{k_0,\ell_0}) \sim \sum_{k=k_0}^{\infty} 2^{-k} \sum_{B_{k,\ell} \subset Q_{k_0,\ell_0}} \tau(B_{k,\ell}) \le \sum_{k=k_0} 2^{-k}(c_k 2^{-k_0} + 1) \le C 2^{-k_0}.$$

We leave it as an exercise to verify that one also can arrange things so that the sequence becomes separated and hence an interpolation sequence. A simple way to do this is instead just to distribute a_j in $B_{k,\ell}$ with positive ℓ when for example k is odd and with negative ℓ when k is even. Furthermore, if the points in each box are appropriately spread out, the sequence will be separated.

We now shall consider another interpretation of condition (4.2). If we write $d\mu(\zeta) = (1-|\zeta|^2)d\nu(\zeta)$, where $d\nu$ is another positive (possibly infinite) measure, then the condition becomes $\int_U K(\zeta, z)d\nu \le C$, where

$$K(\zeta, z) = \frac{(1 - |z|^2)(1 - |\zeta|^2)}{|1 - \bar{\zeta}z|^2}.$$

By the proof of Lemma 3.2, $K(\zeta, z)$ is of the same size as the Green kernel away from the singularity $\zeta = z$. Thus $K(\zeta, z)$ can be seen as a regularization of the Green kernel, and condition (4.2) then says that a regularization of $d\nu$ can be written as the Laplacian of a bounded function. (For measures related to the interpolation theorem, this is the content of condition (3.1).) This point of view leads to a simple proof of the following inequality for Carleson measures.

4.3 Theorem (Carleson's Inequality). *Suppose that μ is a positive measure in U. Then μ is a Carleson measure if and only if*

$$\int_U |f(\zeta)|^p d\mu(\zeta) \le C\|f\|_{H^p}^p, \quad f \in H^p, \ p > 0. \tag{4.3}$$

Clearly, (4.3) holds for all $p > 0$ if it holds for some fixed p since we can divide out the zeros and then take arbitrary p'th roots. In view of Theorem 1.7 in Ch. 6, it also holds for h^p if $p > 1$ (though only with a constant that depends on p). In Chs. 8 and 9 Carleson measures occur, but only as measures satisfying (4.3).

Proof. First suppose that μ satisfies (4.3) for $p = 2$. Since $\zeta \mapsto (1 - \zeta\bar{z})^{-1}$ is analytic for $z \in U$, it follows that (4.2) holds and therefore μ is a Carleson measure according to Proposition 4.1.

Conversely, assume that μ is a Carleson measure. It is enough to show the inequality for say $p = 1$ and $f \in A(U)$ that are smooth up to the boundary. By Proposition 4.1

$$M\mu(z) = -\int_U \frac{1-|z|^2}{|1-\bar{\zeta}z|^2}d\mu(\zeta)$$

is a bounded function, $-C \le M\mu \le 0$, and C^∞ in U. Moreover, a straightforward computation shows that it is subharmonic and, more precisely,

$$\frac{1}{4}\Delta M\mu(z) = \int_U \frac{1-|\zeta|^2}{|1-\bar{\zeta}z|^4}d\mu(\zeta).$$

By formula (2.6) in Ch. 8 (applied to the disks $D(0,r)$ and letting r increase to 1) we get

$$\int_U (1-|z|^2)|f(z)|\Delta M\mu(z)d\lambda(z) \le C\|f\|_{H^1}.$$

In view of Fubini's theorem it is therefore enough to show that

$$|f(\zeta)| \le C \int_U \frac{(1-|\zeta|^2)(1-|z|^2)}{|1-\bar{\zeta}z|^4}|f(z)|d\lambda(z). \qquad (4.4)$$

However, for any $h \in A(U) \cap C^1(\overline{U})$,

$$h(\zeta) = \frac{2}{\pi}\int_U \frac{(1-|z|^2)h(z)d\lambda(z)}{(1-\bar{z}\zeta)^3}, \quad \zeta \in U;$$

see, e.g., Exercise 15 in Ch. 3. Hence (4.4) follows by taking $h(z) = f(z)/(1-\bar{\zeta}z)$. $\qquad\square$

4.4 Remark. The inequality (4.4) is true for any positive subharmonic u instead of $|f|$. In fact, if $\phi(z) = \phi_\zeta(z) = (\zeta - z)/(1 - \bar{z}\zeta)$, then $u \circ \phi^{-1}$ is subharmonic and hence

$$u \circ \phi^{-1}(0) \le C \int_{|w|<\epsilon} u \circ \phi^{-1}(w)d\lambda(w);$$

and so we get that

$$u(\zeta) \le C \int_{|\phi(z)|<\epsilon} \frac{(1-|z|^2)^2}{|1-\bar{\zeta}z|^4}u(z)d\lambda(z).$$

However, in the set $|\phi(z)| < \epsilon$, $1 - |\zeta| \sim 1 - |z|$ and hence (4.4) follows. $\square$

Supplementary Exercises

Exercise 1. Show that $f \in A(U)$ is in $\mathcal{N}$ if and only if $f = g/h$ for some $f, g \in H^\infty$ where h is nonvanishing.

Exercise 2. Show that $\mathcal{N}$ is in fact a linear space.

Exercise 3. Suppose that $f \in A(U)$. Show that $f \in H^p$, $0 < p \leq \infty$, if and only if there is a harmonic u in U such that $|f(z)|^p \leq u(z)$. Also show that if there is some harmonic majorant, then there is at least one u_f. Finally, show that $\|f\|_{H^p} = u_f(0)$.

Exercise 4. Suppose that $f \in A(U)$. Show that $f \in \mathcal{N}$ if and only if $\log^+ |f|$ has a harmonic majorant.

Exercise 5. Show that the polynomials are dense in H^p, $p > 0$.

Exercise 6. Suppose that $f \in A(U)$. Show that f is a Blaschke product if and only if $\lim_{r \nearrow 1} (1/2\pi) \int \left| \log |f(re^{i\theta})| \right| d\theta = 0$.

Exercise 7. Suppose that $f \in A(U)$ and $\operatorname{Re} f > 0$. Show that f is an outer function. Hint: Notice that $\arg f$ is bounded.

Exercise 8. Show that $f(z) = (1-z)^{-1} \log(1-z)$ is in H^p for all $0 < p < 1$ and that f has unbounded Taylor coefficients.

Exercise 9. Suppose that $f \in \mathcal{N}$ and $f^* \in L^1(T)$. Does it follow that $\mathcal{F} f^*(n) = 0$ for $n < 0$? (Answer: No!)

Exercise 10. Suppose that $f \in \mathcal{N}$. Show that $f \in \mathcal{N}^+$ if and only if

$$\lim \frac{1}{2\pi} \int_0^{2\pi} \log^+ |f(re^{i\theta})| d\theta = \frac{1}{2\pi} \int_0^{2\pi} \log^+ |f^*(e^{i\theta})| d\theta.$$

Hint: Compare to Exercise 11 in Ch. 6.

Exercise 11. Suppose that $0 < r < s < \infty$. Show that the inclusions $\mathcal{N} \supset H^r \supset H^s \supset H^\infty$ are proper.

Exercise 12. Suppose that $f \in A(U)$ and that $f(U)$ is not dense in $\mathbb{C}$. Show that $f^* = \lim f_r$ exists for a.e. θ.

Exercise 13. Suppose that K is a compact proper subset of T. Prove that each $f \in C(K)$ can be approximated uniformly by analytic polynomials. Hint: Use the Hahn–Banach theorem.

Exercise 14. Show that if $f \in \mathcal{N}^+$ and $1/f \in \mathcal{N}^+$, then f is an outer function.

Exercise 15. Show that if $f, g \in H^p$ and $|f| \leq |g|$, then $|M_f| \leq |M_g|$ and $|Q_f| \leq |Q_g|$.

Exercise 16. Show that if $f \in \mathcal{N}^+$, then

$$\log |f(0)| \leq \frac{1}{2\pi} \int_0^{2\pi} \log |f^*| dt, \tag{4.5}$$

with equality if and only if M_f is constant.

Exercise 17. Suppose that $f \in \mathcal{N}$ and (4.5) holds. Does it follow that $f \in \mathcal{N}^+$?

Exercise 18. Show Theorem 2.3 without referring to Theorem 2.1: Since trivially $\mathcal{P}(g) \subset M_g H^2$, we have to prove the opposite inclusion. Since $\|M_g f - gp\|_{H^2} = \|f - Q_g p\|_{H^2}$, it is enough to show that $\mathcal{P}(g) = H^2$ if g is outer. If $h \perp \mathcal{P}(g)$, then

$$\frac{1}{2\pi} \int_0^{2\pi} g^* \bar{h}^* e^{-int} dt = 0, \quad n \le 0,$$

and therefore $g^* \bar{h}^* = k^*$ for some $k \in H^1$. However, since g is outer, $k/g \in \mathcal{N}^+$ and $(k/g)^* = \bar{h}^* \in L^2(T)$, and therefore $k/g \in H^2$. It now follows that h^* is constant and hence $h = 0$, since otherwise $g \in H^2$ and $\int g^* = 0$, which would imply that g were divisible by z.

Exercise 19. Show Theorem 2.3 for $1 < p < \infty$ instead of just $p = 2$. Hint: Recall Theorem 2.3 in Ch. 6.

Exercise 20. Show directly that any interpolation sequence must be the zero set of some bounded $f \not\equiv 0$.

Exercise 21. Show that if $\alpha, \beta \in U$ have given modulii, then $|\alpha - \beta|/|1 - \bar{\alpha}\beta|$ is minimal if and only if they lie on the same radius.

Exercise 22. Let T be the operator in the proof of the necessity part of the interpolation theorem. Show that $\operatorname{Ker} T = \{gB; \; g \in H^\infty\}$, where B is the Blaschke product with zeros a_j.

Exercise 23. Show that the sequence $r_k = 1 - 1/k^2$ is not separated. Does it satisfy (3.2)?

Exercise 24. Suppose that a_j is a sequence in U with no limit points and $\beta \in l^\infty$. Show that if there is some $f \in H^\infty$ that interpolates β in a_j, then it is unique if and only if $\sum 1 - |a_j| = \infty$.

Exercise 25. Show (4.1). Hint: Notice that $|1 - r_a e^{-i\theta_1} r_2 e^{i\theta_2}|$ is the Euclidean distance between $r_1 r_2 e^{i(\theta_2 - \theta_1)}$ and the point 1.

Exercise 26. Let μ be a finite measure in U. Show that one can rearrange the mass on each level $\{2^{-(k+1)} \le |z| < 2^{-k}\} = \cup_\ell B_{k,\ell}$ so that the new measure is a Carleson measure.

Exercise 27. Fill in the details in the construction in Example 4.2.

Exercise 28. Let Mf be Hardy–Littlewood's maximal function; see Section 1 in Ch. 6. Use Carleson's inequality to prove that

$$\int_T |Mf^*|^p d\theta \le C\|f\|_{H^p}^p \qquad (4.6)$$

for $p > 0$. The same holds for h^p if $p > 1$. Hint: First prove it for $\tilde{M}$ instead of M. Choose a piecewise C^1 curve $\theta \mapsto r(\theta)e^{i\theta}$ such that

$|f(r(\theta)e^{i\theta})| \geq 1/2\tilde{M}f(e^{i\theta})$. Define a measure on this curve so that the measure of any segment is precisely the length of its projection onto T. Then this measure is a Carleson measure if it is considered as a measure in U. Now Carleson's inequality implies (4.6).

The *hyperbolic distance* $d(z, w)$ between the points z and w in U is defined in the following way:

$$d(0, z) = d(z, 0) = \frac{1}{2} \log \frac{1 + |z|}{1 - |z|},$$

and

$$d(z, w) = d(0, \phi_z(w)).$$

Exercise 29. Show that $d(z, w)$ is well defined for all $z, w \in U$, that $d(z, w) = d(w, z)$, and that

$$d(z, w) = d(\phi(z), \phi(w))$$

for any ϕ that is either an automorphism or the conjugate of some automorphism of U.

A *hyperbolic line* in U is the intersection of U with a circle in $\mathbb{C}$ that intersects T orthogonally. Thus, hyperbolic lines through the origin are just ordinary lines.

Exercise 30. Show that through each disjoint pair of points there is one and only one hyperbolic line.

Exercise 31. Show that all automorphisms and their conjugates map (hyperbolic) lines onto (hyperbolic) lines.

Exercise 32. Show that

$$d(z, w) \leq d(z, z') + d(z', z)$$

with equality if and only if z, z', and w are consecutive points on the same hyperbolic line.

Exercise 33. Show that there are constants c and C such that

$$\frac{1}{C} d(z, w) \leq \left| \frac{z - w}{1 - \bar{z}w} \right| \leq C d(z, w),$$

for all z, w such that $d(z, w) < c$.

The unit disk with this geometry provides a model for non-Euclidean geometry. In fact, all of the Euclidean axioms hold except the axiom of parallels.

Exercise 34. Show that through any point outside some given hyperbolic line there are infinitely many lines that do not intersect the given line.

Notes

We recommend [D] and [G] for further results and references.

The key to the H^p-theory here is Theorem 1.1. There is an analogous result in the upper half-plane. Instead of Theorem 1.1, one can rely on the fact that $|f|^p$ is subharmonic for all $p > 0$. There is a generalization of the H^p-theory to the upper half-space in $\mathbb{R}^{n+1}$ for $p > (n-1)/n$, but then there is no analogue to Theorem 1.1. This theory instead is based on subharmonicity; see [S].

Theorem 1.1 is due to F. Riesz, 1923, and the main theorem (Theorem 1.8) is due to Smirnoff, 1929. See [D] for more historical remarks.

Theorems 2.1 and 2.3 are due to Beurling, 1949. These theorems have been generalized by Gamelin to $p > 0$; see *TAMS*, vol. 124 (1966), 158–167. See also Exercise 19.

The interpolation theorem (Theorem 3.1) was proved by Carleson in 1958. The hard direction is to prove that (3.1) is sufficient. The explicit proof given here was discovered by Jones (*Acta. Math.*, vol. 150 (1983), 137–152).

There is a generalization of the interpolation theorem (Theorem 3.1) to H^p, $p > 0$; see [D].

If the sequence c_k in Example 4.2 satisfies $\sum c_k^2 2^{-k} < \infty$, which is more restrictive than the Blaschke condition, then for almost all choices of θ_k (interpreted in an appropriate sense) $a_k = r_k e^{i\theta_k}$ is an interpolation sequence; see Rudowicz, *Bull. London Math. Soc.*, vol 26 (1994), 160–164.

The proof given here of Carleson's inequality (Theorem 4.3) is due to Berndtsson. The underlying idea that essentially all Carleson measures are of the form $(1 - |\zeta|)\Delta\phi$, where ϕ is a bounded subharmonic function, is from *Proc. Univ. of Maryland 1985-86*, Springer-Verlag (1987). The usual proof of Carleson's inequality proceeds via an $L^p(T)$-estimate for the Hardy–Littlewood maximal function. Incidentally, this estimate follows quite easily from Carleson's inequality; see Exercise 28.

8

Ideals and the Corona Theorem

§1. Ideals in $A(\Omega)$

We begin with

1.1 Theorem. *If* $g_1, \ldots, g_n \in A(\Omega)$ *have no common zeros in* Ω, *then there are* $u_1, \ldots, u_n \in A(\Omega)$ *such that* $\sum_1^n g_j u_j = 1$.

One can obtain Theorem 1.1 quite easily from Weierstrass' and Mittag–Leffler's theorems (see Exercise 1), but we will give a proof that relies on the existence of solutions to the inhomogenous Cauchy–Riemann equation (Theorem 2.3 in Ch. 3). The advantage is that one obtains estimates of the solution $(u_1, \ldots, u_n)$ if one can solve the $\partial / \partial \bar{z}$-equation with estimates. This will be used in the proof of the corona theorem in the next section.

Proof of Theorem 1.1. Consider $g = (g_1, \ldots, g_n)$ as a row matrix and $u = (u_1, \ldots, u_n)^t$ as a column matrix (thus t denotes transpose). We are looking for an analytic u such that $gu = 1$. If $|g|^2 = \sum |g_j|^2$ and

$$\gamma = \frac{\bar{g}^t}{|g|^2},$$

then $\gamma \in C^\infty(\Omega)$ and $g\gamma = 1$. We then want to modify γ to be an analytic solution, and to this end we set

$$u = \gamma + wg^t,$$

where w is an antisymmetric $(n \times n)$-matrix, i.e., $w^t = -w$. Since gwg^t is a scalar (a 1×1-matrix), $gwg^t = (gwg^t)^t = gw^t g^t = -gwg^t$, which gives $gwg^t = 0$, and hence u solves $gu = 1$. If w is now an antisymmetric solution to

$$\frac{\partial w}{\partial \bar{z}} = \left(\left(\frac{\partial \gamma}{\partial \bar{z}} \bar{g} \right)^t - \frac{\partial \gamma}{\partial \bar{z}} \bar{g} \right) \frac{1}{|g|^2} = F \qquad (1.1)$$

$(\partial/\partial\bar{z}$ of course acts componentwise on matrices; observe that the right-hand side in (1.1) is antisymmetric), then

$$\frac{\partial u}{\partial\bar{z}} = \frac{\partial\gamma}{\partial\bar{z}} + \frac{\partial w}{\partial\bar{z}}g^t = \frac{\partial\gamma}{\partial\bar{z}} + \bar{g}^t\left(\frac{\partial\gamma}{\partial\bar{z}}\right)^t g^t \frac{1}{|g|^2} - \frac{\partial\gamma}{\partial\bar{z}}$$

$$= \bar{g}^t\frac{\partial}{\partial\bar{z}}(g\gamma)^t/|g|^2 = \bar{g}^t\frac{\partial 1}{\partial\bar{z}}\frac{1}{|g|^2} = 0,$$

and hence u is in fact an analytic solution as desired. $\Box$

Let

$$[g_1,\ldots,g_n] = \left\{\sum_1^n v_j g_j;\ v_j \in A(\Omega)\right\}$$

denote the ideal in $A(\Omega)$ generated by $g_1,\ldots,g_n$. Theorem 1.1 then says that $[g_1,\ldots,g_n] = [1] = A(\Omega)$ if (and only if) $g_1,\ldots,g_n$ have no common zeros.

1.2 Corollary. *For given $h_1,\ldots,h_n \in A(\Omega)$ there is a $\phi \in A(\Omega)$ such that*

$$[h_1,\ldots,h_n] = [\phi].$$

Thus, each finitely generated ideal in $A(\Omega)$ is a principal ideal.

Proof. According to Weierstrass' theorem there is a $\phi \in A(\Omega)$ that has zeros with right multiplicities exactly where all h_j have common zeros (this is a set with no limit points unless all h_j vanish on some component). However, then the functions $g_j = h_j/\phi$ have no common zeros, and hence by Theorem 1.1 there are u_j such that $\sum u_j h_j/\phi = 1$, i.e., $\sum u_j h_j = \phi$, and therefore $\phi \in [h_1,\ldots,h_n]$. Since $h_j \in [\phi]$ for each j, the corollary is proved. $\Box$

§2. The Corona Theorem

Suppose that $g_1,\ldots,g_h \in H^\infty = H^\infty(U)$ and there are $u_j \in H^\infty$ such that $\sum g_j u_j = 1$. Then by Cauchy–Schwarz' inequality, $\sum |g_j|^2 \geq \delta^2$, where $1/\delta = \sqrt{n}\max_j \|u_j\|_{H^\infty}$. Conversely, we have

2.1 The Corona Theorem. *Suppose that $g_1,\ldots,g_n \in H^\infty$ and there is a $\delta > 0$ such that*

$$\sum_1^n |g_j|^2 \geq \delta^2. \tag{2.1}$$

Then there are $u_1, \ldots, u_n \in H^\infty$ such that

$$\sum g_j u_j = 1.$$

The corona theorem was stated as a conjecture in the 1940s in connection with the study of Banach algebras (this is where the name "corona theorem" originates). It was first proved by Carleson in 1961, and a different proof based on functional analysis was found by Hörmander in 1967. Around 1980 this proof was simplified by Wolff. The proof we give here is a slight modification of Wolff's proof.

To begin with, one can reduce to

2.2 Proposition. *There is $C_{\delta,n} > 0$ such that if $g_1, .., g_n \in H^\infty \cap C^\infty(\overline{U})$ and $\delta \leq |g| \leq 1$, then there are $u_j \in H^\infty$ with $|u| \leq C_{\delta,n}$ such that $\sum g_j u_j = 1$.*

Proof that Proposition 2.2 Implies Theorem 2.1. In fact, we may assume that $\delta < |g| \leq 1$ in the corona theorem. Then $g^r(z) = g(rz)$ satisfies the assumptions in Proposition 2.2, and therefore we get analytic u^r with $|u^r| \leq C_{\delta,n}$ and $g^r u^r = \sum g_j^r u_j^r = 1$. Since $\{u^r\}_{r<1}$ is a normal family (Proposition 1.7 in Ch. 2), we can extract a convergent subsequence $u^{r_j} \to u$ when $r_j \nearrow 1$. Then of course $|u| \leq C_{\delta,n}$ and $gu = 1$, and thus the corona theorem follows.   □

In what follows we therefore may assume that g satisfies the hypotheses in Proposition 2.2. We shall reformulate it as a $\partial/\partial\bar{z}$ problem; but in order to give the precise statement, we need the following formalism. Suppose that $\partial V/\partial\bar{z} = f$, $V \in C^\infty(\overline{U})$, and $v = V|_T$. Then, by Stokes' theorem ((1.5) in Ch. 1),

$$2i \int_U f h d\lambda = \int_T v h dz, \quad \forall h \in A(U) \cap C^\infty(\overline{U}). \tag{2.2}$$

If $f \in L^1(U)$, $v \in L^1(T)$ (or $v \in \mathcal{M}(T)$) and (2.2) holds, one says that v solves the $\partial/\partial\bar{z}$ equation on the boundary and writes

$$\bar{\partial}_b v = f.$$

Note that $\bar{\partial}_b$ is linear, and that $\bar{\partial}_b a v = a f$ if $a \in A(U) \cap C^\infty(\overline{U})$ and $\bar{\partial}_b v = f$. To be precise, $\bar{\partial}_b v$ defines an element in the quotient space

$$L^1(U)/\{g \in L^1(U); \int_U g h d\lambda = 0 \quad \forall h \in A(U) \cap C^\infty(\overline{U})\}.$$

We claim that if, for example, $f \in C^\infty(\overline{U})$ and $\bar{\partial}_b v = f$, then v is the restriction to the boundary of a unique solution V to $\partial V/\partial\bar{z} = f$. In fact, let W be an arbitrary solution in $C^\infty(\overline{U})$ to $\partial W/\partial\bar{z} = f$ (such a solution always exists; however, the reader may just as well assume that f is smooth in a neighborhood of $\overline{U}$; see also Exercise 10). If $w = W|_T$, then $\overline{\partial}_b(v - w) = 0$, which means in particular that $\int_T (v - w)z^n dz = 0$, $n = 0, 1, 2, \ldots$, i.e., all negative Fourier coefficients of $v - w$ vanish; and therefore, $v - w$ is the boundary values of an analytic function $H \in H^1$. However, then v is the boundary values of the function $V = W + H$ that solves $\partial V/\partial \bar{z} = f$.

We shall use the following weak formulation of the $\overline{\partial}_b$ equation.

2.3 Proposition. *Suppose that $f \in C^\infty(\overline{U})$. The equation $\overline{\partial}_b v = f$ has a solution v with $\|v\|_{L^\infty} \leq C$ if and only if*

$$\left| 2 \int_U fh d\lambda \right| \leq C \int_T |h| d\theta, \quad h \in A(U) \cap C^\infty(\overline{U}). \tag{2.3}$$

Proof. Suppose that (2.3) holds; for (the restriction to T of) $h \in A(U) \cap C^\infty(\overline{U})$, let

$$\Lambda h = 2i \int_U fh d\lambda.$$

Then $|\Lambda h| \leq C \int_T |h| d\theta$ and by the Hahn–Banach theorem, Λ can be extended to a bounded linear functional on $L^1(T)$ with norm $\leq C$. Hence there is an $a \in L^\infty(T)$ such that $\|a\|_{L^\infty} \leq C$ and

$$2i \int_U fh d\lambda = \Lambda h = \int_T ha\, d\theta, \quad h \in A(U) \cap C^\infty(\overline{U}).$$

If now $v(z) = -i\bar{z}a(z)$, then (cf. (2.2)) $\overline{\partial}_b v = f$. The converse is trivial and we leave it as an exercise. □

So far we only have made a preliminary reduction (Proposition 2.2) and expressed it in terms of a system of $\overline{\partial}_b$-equations. The next theorem due to Wolff is in fact the heart of the matter.

2.4 Theorem. *There is a $C > 0$ such that if $\phi, f \in C^\infty(\overline{U})$, ϕ real, $|\phi| \leq 1$,*

$$|f|^2 \leq \Delta\phi, \tag{2.4}$$

and

$$\left| \frac{\partial f}{\partial z} \right| \leq \Delta\phi, \tag{2.5}$$

then there is a solution $v \in L^\infty(T)$ to $\overline{\partial}_b v = f$ with $\|v\|_{L^\infty} \leq C$.

Proof of Proposition 2.2. If $\gamma = \bar{g}^t/|g|^2$, then $\gamma \in C^\infty(\overline{U})$, $g\gamma = 1$, and $|\gamma| \leq 1/\delta$. Assume that we can find an (antisymmetric) solution to

$$\overline{\partial}_b w = \left(\left(\frac{\partial \gamma}{\partial \bar{z}} \bar{g} \right)^t - \frac{\partial \gamma}{\partial \bar{z}} \bar{g} \right) \frac{1}{|g|^2} = F$$

such that $w \in L^\infty(T)$ and $|w| \leq C'_{\delta,n}$, where $C'_{\delta,n}$ depends only on δ and n. If $u = \gamma + wg^t$, then (as before) $\bar{\partial}_b u = 0$, $gu = 1$ on T, and $|u| \leq 1/\delta + C'_{\delta,n} = C_{\delta,n}$. However, then u is the boundary values of a $U \in H^1$; and since $\|u\|_{L^\infty} \leq C_{\delta,n}$, U is actually in H^∞ and $\|U\|_{H^\infty} \leq C_{\delta,n}$. Finally, $gU \equiv 1$ since $gU - 1 \in H^\infty$ and the equality holds on T.

Thus, we have to solve $\bar{\partial}_b w = F$ with $L^\infty(T)$-control of the solution. Note that

$$\frac{\partial \gamma}{\partial \bar{z}} = \frac{\partial}{\partial \bar{z}}|g|^{-2}\bar{g} = -2\frac{(g \cdot \bar{g}')\bar{g}}{|g|^4} + \frac{\bar{g}'}{|g|^2},$$

and therefore $|\partial \gamma/\partial \bar{z}| \lesssim |g'|$, which implies that

$$|F|^2 \lesssim |g'|^2 \sim \Delta|g|^2.$$

Here, "$\lesssim$" means "less than or equal to $C_{\delta,n}$ times ... ," where $C_{\delta,n}$ depends only on n and δ. By a similar computation one finds that

$$\left|\frac{\partial F}{\partial z}\right| \lesssim \left|\frac{\partial \gamma}{\partial \bar{z}}\right| |g'| + \left|\frac{\partial^2 \gamma}{\partial z \partial \bar{z}}\right| \lesssim |g'|^2 \sim \Delta|g|^2.$$

Therefore, Proposition 2.2 follows from Theorem 2.4 with $|g|^2 = \phi$.   □

What remains is to prove Theorem 2.4, and to this end we will apply the following simple but useful lemma.

2.5 Lemma. *If $g, \psi \in C^\infty(\overline{U})$, ψ real, and g is analytic, then*

$$\frac{1}{2}\int_U (1 - |z|^2)(\Delta\psi)|g|e^\psi \leq \int_T |g|e^\psi d\theta, \tag{2.6}$$

$$\int_U (1 - |z|^2)(\Delta\psi)|g| \leq 4e\|\psi\|_{L^\infty(U)} \int_T |g|d\theta, \tag{2.7}$$

and

$$\frac{1}{2}\int_U (1 - |z|^2)\frac{|g'|^2}{|g|} \leq \int_T |g|d\theta. \tag{2.8}$$

The inequality (2.7) means that $(1 - |z|^2)\Delta\psi$ is a Carleson measure; cf. Section 4 in Ch. 7.

Proof. First suppose that $g \equiv 1$. Then

$$\frac{1}{4}\int_U (1 - |z|^2)\Delta\psi e^\psi \leq \int_U (1 - |z|^2)\left(\frac{\partial^2\psi}{\partial z \partial \bar{z}} + \left|\frac{\partial\psi}{\partial z}\right|^2\right)e^\psi$$

$$= \int_U (1 - |z|^2)\frac{\partial^2}{\partial z \partial \bar{z}}e^\psi = -\int_U e^\psi + \frac{1}{2}\int_T e^\psi d\theta \leq \frac{1}{2}\int_T e^\psi d\theta,$$

where we have used Green's identity in the last equality. To obtain (2.6) in the general case, note that $\log(|g|^2 + \epsilon)$ is subharmonic, then replace ψ by $\psi + (1/2)\log(|g|^2 + \epsilon)$, and finally let $\epsilon \searrow 0$. One gets (2.7) from (2.6) if ψ is replaced by $\psi/2\|\psi\|_{L^\infty(U)}$. From the computation above it also follows that

$$\int_U (1 - |z|^2)\left|\frac{\partial\psi}{\partial z}\right|^2 e^\psi \leq \frac{1}{2}\int_T e^\psi\, d\theta$$

if $\Delta\psi \geq 0$, and (2.8) follows if one takes $\psi = (1/2)\log(|g|^2 + \epsilon)$ and lets $\epsilon \searrow 0$. □

Proof of Theorem 2.4. In view of Proposition 2.3 we have to estimate $\int_U fh$, and we first rewrite it:

$$\int_U fh = \int_U |z|^2 fh + \int_U (1 - |z|^2) fh$$

and

$$\int_U |z|^2 fh = \int_U fhz\frac{\partial}{\partial z}(|z|^2 - 1) = \int_U (1 - |z|^2)\frac{\partial}{\partial z}(zfh)$$

$$= \int_U (1 - |z|^2)\frac{\partial f}{\partial z}zh + \int_U (1 - |z|^2)f\frac{\partial}{\partial z}(zh),$$

where the second equality follows from Stokes' theorem (the boundary integral vanishes). Thus,

$$\int_U fh = \int_U (1 - |z|^2)\frac{\partial f}{\partial z}zh + \int_U (1 - |z|^2)f\frac{\partial}{\partial z}(zh) + \int_U (1 - |z|^2)fh$$

$$= I_1 + I_2 + I_3,$$

and we have to show that

$$|I_j| \lesssim \int_T |h|, \quad j = 1, 2, 3. \tag{2.9}$$

From (2.7) we immediately get

$$|I_1| \leq \int_U (1 - |z|^2)\Delta\phi|h| \lesssim \int_T |h|.$$

If $g = zh$, we have

$$|I_2|^2 \leq \left(\int (1 - |z|^2)\sqrt{\Delta\phi}|g'|\right)^2$$

$$\leq \int_U (1 - |z|^2)\Delta\phi|g| \int_U (1 - |z|^2)\frac{|g'|^2}{|g|} \lesssim \left(\int_T |g|\right)^2 = \left(\int_T |h|\right)^2$$

by (2.7) and (2.8), and therefore (2.9) holds for $j = 2$. The estimation of I_3 is left as an exercise. □

2.6 Remark. Wolff's original formulation of Theorem 2.4 reads: *Suppose that the Carleson norms of* $(1-|z|)|f|^2$ *and* $(1-|z|)|\partial f/\partial z|$ *are* ≤ 1. *Then there is a solution* $v \in L^\infty(T)$ *to* $\bar\partial_b v = f$ *with* $\|v\|_{L^\infty} \leq C$, *where* C *is an absolute constant.* See Exercise 19. The discussion preceding Theorem 4.3 motivates the relation between the two formulations. Notice that, contrary to the case of the interpolation theorem, the Carleson measures that appear in the proof of the corona theorem are already smooth, and therefore we do not need to worry about regularizations. See also Exercises 20 and 22.

In this context it is natural to mention that there is a very general result due to Hörmander about L^2-estimates of solutions to the inhomogeneous Cauchy–Riemann equation(s) in several complex variables. Here is the one-variable case. A proof is outlined in Exercises 23 and 24.

2.7 Theorem. *Suppose that* $\Omega \subset \mathbb{C}$ *is bounded and* ψ *is any subharmonic function in* Ω. *For any smooth function* f *there is a (smooth) solution* u *to* $\partial u/\partial \bar z = f$ *such that*

$$\int_\Omega |u|^2 e^{-\psi} \leq C^2 \int_\Omega |f|^2 e^{-\psi},$$

where C *is the diameter of* Ω.

Supplementary Exercises

Exercise 1. Here is an outline of an alternative proof of Theorem 1.1: Suppose that it is true for $n-1$ generators. Take a ϕ that has exactly the zeros that are common to $g_1, \ldots, g_{n-1}$. Then by the induction hypothesis there are v_j such that $g_1 v_1 + \ldots g_{n-1} v_{n-1} = \phi$. Since g_n is nonvanishing at the zeros of ϕ, there is by Corollary 1.6 to Weierstrass' theorem a h such that $(1 - g_n h)/\phi = f$ is analytic. Hence, $1 = \phi f - g_n h = g_1 f v_1 + \ldots g_{n-1} f v_{n-1} - g_n h$.

Exercise 2. For $N = 1, 2, 3, \ldots$, let

$$g_N(z) = \prod_{n=N}^{\infty} \left(1 - \frac{z^2}{n^2}\right).$$

Show that the ideal in the ring of entire functions, generated by $\{g_N\}$, is not a principal ideal.

Exercise 3. Show that the solution $\gamma = \bar g^t/|g|^2$ is the solution to $g\gamma = 1$ that has the pointwise minimal Euclidean norm.

Exercise 4. Suppose that μ is a measure on T. Show that $\mu = f d\theta$ for some $f \in L^2(T)$ if and only if $(1 - |\zeta|^2)|\nabla P\mu(\zeta)|^2$ is integrable over U.

Exercise 5. Suppose that f is C^∞ in a neighborhood of $\overline{U}$, $\mu \in \mathcal{M}(T)$, and $\overline{\partial}_b \mu = f$. Show that μ is absolutely continuous.

Exercise 6. Prove the "only if" part of Proposition 2.3.

Exercise 7. Suppose that $h, g_1, \ldots, g_n \in H^\infty$ and $|h| \leq |g|^3$. Show that there are $u_1, \ldots, u_n \in H^\infty$ such that $\sum g_j u_j = h$. Hint: Copy the proof of the corona theorem.

Exercise 8. Show that 3 in the preceding exercise may be replaced by $2 + \epsilon$. First show that

$$|g'|^2/g^{2-2\epsilon} \leq C_\epsilon \Delta |g|^{2\epsilon}.$$

Exercise 9. Suppose that $g_j \in H^\infty$, $j = 1, \ldots, n$ and $\sum |g_j|^2 \geq \delta^2$. Let

$$\psi(\zeta, z) = \left(\frac{1 - |\zeta|^2}{1 - \overline{\zeta}z}\right)^\alpha \left(\frac{\sum g_j(z)\overline{g_j(\zeta)}}{\sum |g_j(\zeta)|^2}\right)^2, \quad \alpha > 0.$$

Show that if $\phi \in A(U) \cap C^\infty(\overline{U})$, then

$$\phi(z) = -\frac{1}{\pi} \int_U \phi(\zeta) \frac{\partial \psi / \partial \overline{\zeta}(\zeta, z)}{\zeta - z} d\lambda(\zeta),$$

and that $\partial \psi / \partial \overline{\zeta}(\zeta, z)/(\zeta - z) \in C^\infty(\overline{U \times U})$. Give explicit $u_j = T_j \phi \in A(U) \cap C^\infty(\overline{U})$ such that $\sum g_j u_j = \phi$. Show that $T_j \phi \in H^p$ if $\phi \in H^p$, $1 \leq p < \infty$.

Exercise 10. Show that

$$u(z) = -\frac{1}{\pi} \int_U \frac{f(\zeta)d\lambda(\zeta)}{\zeta - z}$$

is a $C^\infty(\overline{U})$-solution to $\partial u / \partial \overline{z} = f$ if $f \in C^\infty(\overline{U})$. Hint: First show that if there is some solution in $C^\infty(\overline{U})$ or at least in $C^k(\overline{U})$, then $u(z)$ is in $C^{k-1}(\overline{U})$. To this end use Cauchy's formula 1.6 in Ch. 1. (Since f has some C^k-extension (in fact a C^∞-extension) to a neighborhood of $\overline{U}$, it also has a C^k-solution on $\overline{U}$.)

Alternatively, one can verify the formula

$$\frac{\partial^m u}{\partial z^m} = -\frac{m}{\pi} \int_U \left(\frac{1 - |\zeta|^2}{1 - \overline{\zeta}z}\right)^m \frac{(\partial^m f / \partial \zeta^m)(\zeta)d\lambda(\zeta)}{\zeta - z}.$$

Exercise 11. Show the corona theorem in $A = \{1 < |z| < 2\}$. Hint: Note that the corona theorem is true in any simply connected domain, in particular in $D = \{0 < \operatorname{Re} z < \log 2\}$. There is a one-to-one correspondence between functions in A and functions in D such that $g(z + m2\pi i) = g(z)$. If $f_j \in H^\infty(D)$ are periodic in this way and $\sum |f_j|^2 \geq \delta^2$, show that there are *periodic* $u_j \in H^\infty(D)$ such that $\sum f_j u_j = 1$.

Exercise 12. Show that if $f \in C^\infty(\overline{U})$, then

$$u(z) = \frac{1}{\pi} \int_U \frac{\bar{z}f(\zeta)d\lambda(\zeta)}{1 - \bar{\zeta}z}, \quad z \in T,$$

is a solution to $\bar{\partial}_b u = f$.

Exercise 13. Show that if $\phi \in H^p$, $1 < p < \infty$, and f satisfies (2.4) and (2.5), then

$$v(z) = \frac{1}{\pi} \int_U \frac{\bar{z}f(\zeta)\phi(\zeta)d\lambda(\zeta)}{1 - \bar{\zeta}z}, \quad z \in T,$$

is a solution to $\bar{\partial}_b u = f\phi$ and

$$\int_T |v|^p \leq C_p \int_T |\phi|^p.$$

What happens for $p = 1$?

Exercise 14. Prove (2.9) for $j = 3$. Show that

$$\int_U (1 - |z|^2)|h| \leq C \int_T |h|d\theta$$

for holomorphic h.

Exercise 15. Suppose that $f \in H^\infty$. Show that $(1 - |z|)|f'|^2$ is a Carleson measure.

Exercise 16. Assume that $g \in H^2$. Show that

$$2 \int_U (1 - |z|^2)|g'|^2 \leq \int_T |g|^2 d\theta. \tag{2.10}$$

Exercise 17. Show that for the proof of Theorem 2.4 all one needs is (2.10). Note that $\Delta\phi$ is actually only (a constant times) $|g'|^2$ for a (sum of) bounded analytic functions.

Exercise 18. Suppose that ψ is bounded and subharmonic in U. Show that $\mu = (1 - |\zeta|)\Delta\psi d\lambda(\zeta)$ is a Carleson measure. Hint: Cf. (2.7).

Exercise 19. Prove Wolff's theorem as it is stated in Remark 2.6. Hint: Use Theorem 4.3 in Ch. 7.

Exercise 20. Suppose that μ is a Carleson measure in U. Prove that there is a solution $v \in L^\infty(T)$ to $\bar{\partial}_b v = \mu$.

Exercise 21. Here is an outline of a proof of Carleson's interpolation theorem (Theorem 3.1 in Ch. 7) based on the $\bar{\partial}_b$ equation: The separation hypothesis ensures that we can find $\chi_j \in C_0^\infty(U)$ with mutually disjoint supports such that $\chi_j = 1$ when $d(z, a_j) < \epsilon$ and vanish when $d(z, a_j) > 2\epsilon$ $(d(z, w) \sim |z - w|/|1 - \bar{z}w|)$. Then $\phi = \sum \beta_j \chi_j$ is a smooth solution to the interpolation problem. Let B be the Blaschke product with zeros at a_j. Then

$$F = \sum_j \beta_j \frac{\partial \chi_j}{\partial \bar{z}} / B$$

is a Carleson measure (note that $\int |\partial \chi_j / \partial \bar{\zeta}| \leq C(1 - |a_j|)$). Solve $\bar{\partial}_b v = F$ and take g such that $g|_T = \phi - v$. One can in fact assume that there are only a finite number of points a_j since the occurring constants will be uniform. The general case then follows by a normal family argument.

Exercise 22. In Carleson's original proof of the corona theorem one of the main steps was to find a C^∞ solution γ to $g\gamma = 1$ such that $|\partial \gamma / \partial \bar{z}|$ is a Carleson measure. Given such a γ, complete the proof of the corona theorem.

Exercise 23. If $\psi \in C^2(\Omega)$ is real and $\Delta \psi > 0$, then

$$\frac{\partial u}{\partial \bar{z}} = f \qquad (2.11)$$

has a solution (in the distribution sense) such that

$$\int_\Omega |u|^2 e^{-\psi} \leq 4 \int_\Omega |f|^2 e^{-\psi} / \Delta \psi$$

provided the right-hand side is finite. Sketch of proof: Let $(g, \chi) = \int_\Omega g\bar{\chi} e^{-\psi}$, and let

$$\vartheta = -\frac{\partial}{\partial z} + \frac{\partial \psi}{\partial z}.$$

Then $(\partial g / \partial \bar{\zeta}, \chi) = (g, \vartheta \chi)$ for $\chi \in C_0^\infty(\Omega)$. Show that (2.11) has a solution with $(u, u) \leq C^2$ if and only if $|(f, \chi)|^2 \leq C^2(\vartheta \chi, \vartheta \chi)$ for $\chi \in C_0^\infty(\Omega)$. Verify the equality

$$-\vartheta \frac{\partial}{\partial \bar{\zeta}} + \frac{\partial}{\partial \bar{\zeta}} \vartheta = \frac{\partial^2}{\partial \zeta \partial \bar{\zeta}} \psi = \frac{1}{4} \Delta \psi.$$

Use it to prove that $|(f, \chi)|^2 \leq 4(\vartheta \chi, \vartheta \chi)(f/(\Delta \psi), f)$ and conclude the theorem.

Exercise 24. Use the results in the preceding exercise to prove Theorem 2.7.

Exercise 25. Let $A = A(U) \cap C(\bar{U})$, the so-called *disk algebra*. Suppose that $g_1, \ldots, g_m \in A$ and $\sum |g_j| > 0$. Show that there are $u_j \in A$ such that $\sum g_j u_j = 1$. Hint: First, construct a solution γ_j to $\sum g_j \gamma_j = 1$ such that $\gamma_j, \partial \gamma_j / \partial \bar{z} \in C(\bar{U})$. To this end, take preliminarily γ_j as $1/g_j$ times the characteristic function of the set $E_j = \{z \in \bar{U}; |g_j(z)| > 0\}$ and use a smooth partition of unity. Then proceed as in the proof of Theorem 1.1.

Notes

A thorough discussion of the history of the corona problem and various aspects of proofs are given in [G]. The best known constant (in Proposition 2.2), due to Uchiyama, is $C_\delta = \mathcal{O}(\delta^{-2} \log \delta)$, which is independent of the number of generators n. For the proof and various vector-valued versions of the corona theorem, see [N] Appendix 3, and the references given there.

The corona theorem is closely related to the interpolation theorem (Theorem 3.1 in Ch. 7). Both can be reduced to the problem of finding a bounded solution to a certain $\overline{\partial}_b$-problem. In the case of the interpolation theorem, the occurring right-hand side is a Carleson measure (cf. Exercise 21) and is therefore readily solved by Carleson's inequality (Theorem 4.3 in Ch. 7) and Proposition 2.3. However, in the case of the corona problem, the most obvious right-hand side occurring from the smooth solution $\gamma = \overline{g}^t/|g|^2$ requires the more involved Theorem 2.4 to get a bounded solution. However, in Carleson's original proof a more refined γ occurs such that $\partial\gamma/\partial\overline{z}$ is actually a Carleson measure. By this choice of γ the corona theorem readily follows; cf. Exercise 22.

In the paper by Jones referred to in the notes to the previous chapter, an explicit bounded solution to $\overline{\partial}_b u = f$ is constructed, provided f is a Carleson measure. No such formula is known if f satisfies the hypothesis in Wolff's theorem.

The corona theorem holds in, e.g., multiply connected domains (a finite number of holes), but the general case is not known; cf. Exercise 11.

The analogue question for $A = A(U) \cap C(\overline{U})$ is true, i.e., if $g_j \in A$ has no common zeros, then there are $u_j \in A$ such that $\sum g_j u_j = 1$. This is a simple result in the theory of Banach algebras; see, e.g., Ch. 18 in [Ru1]. See also Exercise 25, where another proof is suggested.

Hörmander's theorem (of which Theorem 2.7 is a special case) was published in *Acta Math.*, vol 113 (1965). This is of particular importance in the theory of several complex variables.

If the power 3 in Exercise 7 is replaced by 1, the statement is false. The case with power 2 is unknown; see [G] and [N].

9

H^1 and BMO

§1. Bounded Mean Oscillation

Recall from Ch. 6 that a function u on T is in $L^p(T)$ if and only if its Hilbert transform Hu is. By virtue of (2.8) in Ch. 6, we can define the Hilbert transform even for $u \in L^1(T)$ as a formal Fourier series; in general, it will not belong to $L^1(T)$ but merely be a distribution; cf. Remark 2.1 in Ch. 6. More precisely, Hu is in $L^1(T)$ if and only if $u = \operatorname{Re} f^*$ for some $f \in H^1$.

Definition. $H^1_{\mathbb{R}}$ is the space of (complex-valued) functions in $u \in L^1(T)$ such that also $Hu \in L^1(T)$ with norm

$$\|u\|_{H^1_{\mathbb{R}}} = \|u\|_{L^1} + \|Hu\|_{L^1}.$$

Some observations are immediate:
(i) Since H is a real opererator, u is in $H^1_{\mathbb{R}}$ if and only if both $\operatorname{Re} u$ and $\operatorname{Im} u$ are.
(ii) The Hilbert transform H maps $H^1_{\mathbb{R}}$ into itself.
(iii) There are strict inclusions

$$L^1(T) \supset H^1_{\mathbb{R}} \supset L^p(T) \qquad \text{if} \quad p > 1. \tag{1.1}$$

(iv) If u is real, then $u \in H^1_{\mathbb{R}}$ if and only if $u = \operatorname{Re} f^*$ for some $f \in H^1$ and there is a constant $C > 0$ such that

$$\frac{1}{C}\|u\|_{H^1_{\mathbb{R}}} \le \|f\|_{H^1} \le C\|u\|_{H^1_{\mathbb{R}}} \tag{1.2}$$

if $f(0)$ is real.
(v) If functions in H^1 are identified by their boundary values, then H^1 is a subspace of $H^1_{\mathbb{R}}$ and the Szegö projection maps $S\colon H^1_{\mathbb{R}} \to H^1$.

Thus, $H^1_{\mathbb{R}}$ is a somewhat smaller space than $L^1(T)$, but with the advantage that the Hilbert and Szegö transforms are bounded. In some respect $H^1_{\mathbb{R}}$ therefore serves as an approriate surrogate for $L^1(T)$. Because of the inclusions (1.1) the dual space of $H^1_{\mathbb{R}}$ has to be something between $L^\infty(T)$ and $L^p(T)$, and the main objective in this chapter is to determine this dual space. It should be pointed out that the $H^1_{\mathbb{R}}$ condition is not a simple condition on the size of $|f|$. In fact, if $|f|$ is in $H^1_{\mathbb{R}}$, then f also is, but the converse is not true; see Exercises 6 and 7.

1.1 Proposition. $H^1_{\mathbb{R}}$ *is a Banach space and $C^\infty(T)$ is a dense subspace.*

Proof. It is enough to consider the subspace of real functions. By (1.2) this subspace is isomorphic to $\{f \in H^1;\ f(0) \text{ is real}\}$, which is clearly a Banach space. Moreover, any f in this space can be approximated by, e.g., the C^∞-functions $f_r(e^{it}) = f(re^{it})$ in H^1 norm; cf. Theorem 1.2 in Ch. 7.□

If I is an interval on T and u is an integrable function, let u_I denote the mean value over I,

$$u_I = \frac{1}{|I|} \int_I u(t)dt,$$

where $|I|$ is the length of I (for simplicity we often write $u(t)$ rather than $u(e^{it})$).

Definition. BMO is the space of all functions $u \in L^2(T)$ such that

$$\|u\|_*^2 = \sup_I \frac{1}{|I|} \int_I |u(t) - u_I|^2 dt + \frac{1}{2\pi} \int_0^{2\pi} |u(t)|^2 dt \qquad (1.3)$$

is finite, where the supremum runs over all intervals I on the unit circle.

It is readily verified that $\|\ \|_*$ is a norm and BMO actually is a Banach space; in fact, any Cauchy sequence in BMO must have a limit at least in $L^2(T)$, and it is easy to see that this sequence actually converges to this limit in the BMO norm. (Sometimes the last term in (1.3) is omitted in the definition. Then $\|\ \|_*$ is just a seminorm since it is independent of additive constants, and BMO then becomes a space of functions modulo constants.) Moreover,

$$\|u\|_* \leq \sqrt{5}\|u\|_{L^\infty},$$

and therefore BMO contains $L^\infty(T)$. However, the inclusion is proper; one can show directly from the definition that, e.g., $\log|t|$ (or $\log|1 - e^{it}|$) is in BMO, whereas on the other hand $(t/|t|)\log|t|$ is not; see Example 1.5 below and Exercise 2. However, if f belongs to BMO, then $|f|$ also does (Exercise 3).

Exercise 1. For $u \in L^2(T)$ let $u_r(e^{it}) = Pu(re^{it})$, where Pu is the Poisson integral of u. Prove that $\|u_r\|_* \to \|u\|_*$ when $r \nearrow 1$ if $u \in BMO$.

1.2 Remark. *BMO* stands for *bounded mean oscillation*, and in our definition we have measured this mean oscillation in the $L^2(T)$-norm. It is more common to define *BMO* with $L^1(T)$-norms instead. This definition is as follows: An $f \in L^1(T)$ is in *BMO* if

$$\sup_I \frac{1}{|I|} \int_I |f - f_I| dt + \frac{1}{2\pi} \int_0^{2\pi} |f| dt$$

is finite. It is trivial that f satisfies this definition if it is in *BMO* in our sense. The converse is also true and is a consequence of the John–Nirenberg theorem, and so these two definitions are equivalent. The reason to use L^2 instead of L^1 is merely practical; it simplifies the proofs of the main results.

The main result in this chapter is Fefferman's theorem, which states that *BMO* is the dual of $H^1_{\mathbb{R}}$ (Theorem 2.1). However, we first shall consider some useful alternative ways to express the *BMO* norm (Proposition 1.3 below). To this end we need the Riesz decomposition in its simplest form for functions that are smooth up to the boundary; cf. (2.6) in Ch. 4: If $\xi \in C^\infty(\overline{U})$, then

$$\xi(z) = P\xi(z) + \mathcal{G}(\Delta\xi)(z), \quad z \in U, \tag{1.4}$$

where P is the Poisson integral and $\mathcal{G}$ the Green potential.

1.3 Proposition. *Suppose that* $u \in L^2(T)$. *Then*

$$\sup_I \frac{1}{|I|} \int_I |u - u_I|^2 dt$$
$$\sim \sup_{a \in U} P(|u - Pu(a)|^2)(a) = \sup_{a \in U} -\mathcal{G}(|\nabla Pu|^2)(a). \tag{1.5}$$

In particular, all of them are finite if one of them is.

1.4 Corollary. *The Hilbert transform maps* BMO *into* BMO *boundedly.*

Proof. Let $u \in L^2(T)$. Since $Pu + iPHu$ is analytic,

$$(\partial/\partial\bar{z})Pu = -i(\partial/\partial\bar{z})PHu;$$

so

$$|\nabla PHu|^2 = |\nabla Pu|^2 \tag{1.6}$$

for real u. The complex case then follows. Hence, by Proposition 1.3, u is in *BMO* if and only Hu is.   □

1.5 Remark. The function $v(t) = \log|1 - e^{it}|$ is in *BMO* since it is the Hilbert transform of the bounded function $\arg(1 - e^{it})$.

Proof of Proposition 1.3. First suppose that $u \in C^\infty(T)$. Then $Pu \in C^\infty(\overline{U})$ and we therefore can apply the Riesz decomposition (1.4) to $\xi = |Pu - Pu(a)|^2$. If we evaluate at the point $z = a$, we get that

$$-\mathcal{G}(|\nabla Pu|^2)(a) = P(|u - Pu(a)|^2)(a),$$

which shows the second part of (1.5) for smooth u. For the general case we approximate u with u_r as in Exercise 1. Then it is enough to verify that for fixed $a \in U$,

$$P(|u_r - Pu_r(a)|^2)(a) \to P(|u - Pu(a)|^2)(a)$$

and

$$\mathcal{G}(|\nabla Pu_r|^2)(a) \to \mathcal{G}(|\nabla Pu|^2)(a)$$

when $r \nearrow 1$. The first limit is quite trivial, and so we concentrate on the second one. Since the Green kernel is $\mathcal{O}(1 - |\zeta|)$ for fixed $z \in U$, it is enough to verify that

$$\int_U (1 - |\zeta|)|\nabla Pu_r|^2 \to \int_U (1 - |\zeta|)|\nabla Pu|^2.$$

Noting that $Pu_r(\zeta) = Pu(r\zeta)$ and therefore $\nabla Pu_r(\zeta) = r\nabla Pu(r\zeta)$, this follows by a substitution of variables and monotone convergence.

We now prove that the first expression in (1.5) is dominated by the second one. By rotation symmetry it is enough to consider intervals I, centered at the origin. If χ_ρ denotes the characteristic function of the interval $(-\rho, \rho)$, then there is a constant C, independent of ρ, such that

$$\frac{1}{2\rho}\chi_\rho(t) \le CP_r(t),$$

where $P_r(t)$ is the Poisson kernel and $r = 1 - \rho$. Thus,

$$\frac{1}{2\rho}\int_{-\rho}^{\rho} |u(e^{it}) - Pu(r)|^2 dt \le CP(|u - Pu(r)|^2)(r),$$

and now the $\lesssim$ part of $\sim$ follows by

1.6 Lemma. *If $u \in L^2(T)$ and to each interval I we have a number a_I, then*

$$\frac{1}{4}\sup_I \frac{1}{|I|}\int_I |u - u_I|^2 \le \sup_I \frac{1}{|I|}\int_I |u - a_I|^2.$$

Proof. Denote the right-hand side by A^2. Then

$$\sqrt{\frac{1}{|I|}\int_I |u - u_I|^2} \leq \sqrt{\frac{1}{|I|}\int_I |u - a_I|^2} + \sqrt{\frac{1}{|I|}\int_I |u_I - a_I|^2}$$

$$\leq A + |u_I - a_I| \leq A + \sqrt{\frac{1}{|I|}\int_I |u - a_I|^2} \leq 2A.$$

In the next to last inequality we have applied Jensen's inequality. $\quad\square$

It remains to show the $\gtrsim$ part of (1.5). By the variant of Lemma 1.6 where the Poisson kernel and the χ_ρ are interchanged, it is enough to find for each $a \in U$ a number a' such that $\sup_{a \in U} P(|u - a'|^2)(a)$ is dominated by the left-most expression in (1.5). We may assume that $a = r > 0$. Then we choose $a' = u_\rho = u_{(-\rho,\rho)}$ (again $r = 1 - \rho$). Now

$$P(|u - u_\rho|^2)(r) = \int_{|t|\leq\rho} P_r(t)|u(t) - u_\rho|^2 dt + \int_{\rho\leq|t|\leq\pi} P_r(t)|u(t) - u_\rho|^2 dt.$$

The first of these integrals is easily estimated by the expression on the left in (1.5) since

$$P_r(t) \leq C\frac{1}{2\rho}\chi_\rho(t)$$

if $|t| \leq \rho$. For the last one, first note that

$$P_r(t) = \frac{1 - r^2}{|1 - re^{it}|^2} \lesssim \frac{\rho}{\rho^2 + t^2} \lesssim \frac{\rho}{t^2}$$

if $|t| \geq \rho$. Thus, the proof is concluded with

1.7 Lemma. *If $u \in L^2$, then*

$$\int_{\rho\leq|t|\leq\pi} \frac{\rho}{t^2}|u(t) - u_\rho|^2 dt \leq C\sup_I \frac{1}{|I|}\int_I |u - u_I|^2 = CB^2,$$

where C is an absolute constant.

Proof of Lemma 1.7. Note that

$$|u_{2^{k-1}\rho} - u_{2^k\rho}|^2 = \left|\frac{1}{2^{k-1}\rho}\int_{|t|\leq 2^{k-1}\rho} (u - u_{2^k\rho})\right|^2$$

$$\leq \frac{1}{2^{k-1}\rho}\int_{|t|\leq 2^k\rho} |u - u_{2^k\rho}|^2 \leq 2B^2,$$

so that $|u_{2^{k-1}\rho} - u_{2^k\rho}| \leq \sqrt{2}B$. Therefore,

$$|u_{2^k\rho} - u_\rho| \leq k\sqrt{2}B$$

and hence

$$\int_{|t|\leq 2^k \rho} |u - u_\rho|^2 \leq 2^k \rho 2(1 + 2k^2)B^2. \tag{1.7}$$

(Here we have the convention that the integration is performed over the whole unit circle if $2^k \rho > \pi$.) Hence,

$$\int_{\rho \leq |t| \leq \pi} \frac{\rho}{t^2} |u(t) - u_\rho|^2 dt \leq \sum_1^\infty \int_{2^{k-1}\rho \leq |t| \leq 2^k \rho} \frac{\rho}{t^2} |u(t) - u_\rho|^2 dt$$

$$\leq \sum_1^\infty \frac{\rho 2(1 + 2k^2)2^k \rho}{(2^{k-1}\rho)^2} B^2 \sim B^2,$$

and thus the lemma is proved. $\quad\square$

As noted above, Proposition 1.3 now is also proved. $\quad\square$

§2 The Duality of H^1 and BMO

Now we are prepared for the main theorem. In this chapter the notion of a Carleson measure refers to a positive measure in U that satisfies Carleson's inequality (4.3) in Ch. 7.

2.1 Theorem (Fefferman). *Suppose that $u \in L^2(T)$. Then the following four statements are equivalent:*
(i) *$u \in BMO$.*
(ii) *$(1 - |\zeta|^2)|\nabla Pu|^2$ is a Carleson measure in U.*
(iii) *u operates on $H^1_{\mathbb{R}}$ in the sense that the functional*

$$\phi \mapsto (\phi, u) = \frac{1}{2\pi} \int_0^{2\pi} \bar{u}\phi dt$$

has a continuous extension from $C^\infty(T)$ to $H^1_{\mathbb{R}}$.
(iv) *$u = a + Hb$ for some $a, b \in L^\infty(T)$.*
    The BMO-norm of u is equivalent to the functional norm in (iii), and to the square root of the Carleson norm in (ii) plus the $L^2(T)$-norm of u.

Before the proof we consider a few consequences and corollaries of this theorem. The equivalence of (i) and (iii) can be rephrased as

$$(H^1_{\mathbb{R}})^* = BMO.$$

In fact, (i) $\to$ (iii) means that any $u \in BMO$ corresponds to a (unique) functional on $H^1_{\mathbb{R}}$ and any functional on $H^1_{\mathbb{R}}$ is in particular a functional on $L^2(T)$, and so it is given by some $u \in L^2(T)$. However, then (iii) $\to$ (i) means that actually $u \in BMO$.

It also follows that we have a continuous inclusion

$$BMO \subset L^p(T)$$

for all $p < \infty$.

Definition. We let $BMOA$ denote the space of analytic functions in H^2 such that their boundary values are in BMO.

As we sometimes consider H^p ($1 \le p \le \infty$) to be a subspace of $L^p(T)$, it is reasonable to think of $BMOA$ as a subspace of BMO. Since the Hilbert transform is bounded on BMO, it follows that the Szegö projection is also. In view of Theorem 2.1, any $b \in BMOA$ operates on H^1 via the pairing (f, b), a priori defined for $f \in H^2$, and we get the following extension of Theorem 2.3 in Ch. 6 (by a similar argument).

2.2 Theorem.

(a) *The Szegö projection is a bounded map*

$$S: BMO \to BMOA \qquad \text{and} \qquad S: H^1_{\mathbb{R}} \to H^1.$$

(b) *$BMOA$ operates on H^1 in the sense that the pairing (f, b) has a continuous extension from H^2 to H^1 if $b \in BMOA$. Conversely, any bounded functional on H^1 occurs in this way, and the functional norm is equivalent to the $BMOA$ norm of b.*

Proof of Theorem 2.1. (i) → (ii): By Proposition 1.3

$$C^2 = \sup -\mathcal{G}(|\nabla Pu|^2) \lesssim \|u\|_*^2.$$

Recall that $\Delta \mathcal{G}\chi = \chi$ if χ is smooth up to the boundary. If f is analytic, then (2.7) in Ch. 8 gives (first replace u by u_r and then take limits)

$$\int_U |f|(1 - |\zeta|^2)|\nabla Pu|^2 = \int_U |f|(1 - |\zeta|^2)\Delta \mathcal{G}(|\nabla Pu|^2) \lesssim C^2 \int_T |f| d\theta.$$

(ii) → (iii): It is enough to show that there is a constant such that

$$|(\phi, u)| \le C\|\phi\|_{H^1_{\mathbb{R}}}\|u\|_*$$

for real $u, \phi \in C^\infty(T)$. Then the inequality follows for any $b \in BMO$ by an approximation as in Exercise 1. By Jensen's formula (2.7) in Ch. 4,

$$(u, \phi) = \frac{1}{2\pi} \int_0^{2\pi} u\phi dt = \frac{1}{\pi} \int_U \log \frac{1}{|\zeta|} \nabla Pu \cdot \nabla P\phi d\lambda(\zeta) + Pu(0)P\phi(0).$$

Since $-\log|\zeta| \sim 1 - |\zeta|$ for $1/2 \le |\zeta| < 1$, we get that

$$|(\phi, u)| \le \int_U (1 - |\zeta|)|\nabla Pu||\nabla P\phi| d\lambda + C \sup_{|\zeta| \le 1/2} |Pu||P\phi|. \qquad (2.1)$$

However,

$$\sup_{|\varsigma|\leq 1/2} |Pu||P\phi| \lesssim \|u\|_{L^2}\|\phi\|_{L^2} \lesssim \|\phi\|_{H^1_{\mathbb{R}}}\|u\|_*,$$

and so it remains to estimate the integral in (2.1). Let f be the unique element in H^1 such that $\operatorname{Im} f(0) = 0$ and $\operatorname{Re} f = P\phi$. Then $\|u\|_{H^1_{\mathbb{R}}} \sim \|f\|_{H^1}$ and $|\nabla P\phi| \sim |f'|$. Thus, the square of the integral is

$$\lesssim \left(\int_U (1 - |\varsigma|)|f'||\nabla Pu| \right)^2$$

$$\lesssim \int_U (1 - |\varsigma|)\frac{|f'|^2}{|f|} \int_U (1 - |\varsigma|)|f||\nabla Pu|^2 \lesssim \left(\int_T |f| d\theta \right)^2,$$

where the last inequality follows from (2.8) in Ch. 8 and the assumption (ii).

We now prove that (iii) implies (iv). If u satisfies (iii), then it defines a bounded functional Λ on $H^1_{\mathbb{R}}$, but $H^1_{\mathbb{R}}$ can be isometrically identified with the subspace of $L^1(T) \times L^1(T)$ of elements of the form $\langle \phi, H\phi \rangle$; and so by the Hahn–Banach theorem, Λ can be extended to a bounded functional on $L^1(T) \times L^1(T)$. Thus, it is represented by an element $\langle a, -b \rangle \in L^\infty(T) \times L^\infty(T)$, and for all $\langle \phi, H\phi \rangle$ we thus have

$$(\phi, u) = \Lambda\phi = (\phi, a) + (H\phi, -b) = (\phi, a + Hb)$$

for smooth ϕ (cf. formula (2.9) in Ch. 6), and hence $u = a + Hb$. Also note that $\|a\|_{L^\infty} + \|b\|_{L^\infty}$ is equivalent to the functional norm of u.

The implication (iv) → (i) follows from Corollary 1.4, and hence Theorem 2.1 is completely proved. □

Supplementary Exercises

Exercise 2. Show directly from the definition that $\log |t|$ is in BMO but that $(t/|t|) \log |t|$ is not. Also show that $\log |1 - e^{it}|$ is in BMO.

Exercise 3. Show that $\||f\||_* \leq 2\|f\|_*$.

Exercise 4. Suppose that $f \in L^1(T)$ (or that f is in $\mathcal{M}(T)$). Show that if $Hf \in \mathcal{M}(T)$, then $f \in H^1_{\mathbb{R}}$.

Exercise 5. Suppose that $f \in L^1(T)$ (or that f is any distribution on T). Show that $f \in H^1_{\mathbb{R}}$ if and only if $|\int f\bar{b}| \leq C\|b\|_*$ for all $b \in C^\infty(T)$. Hint: Use the preceding exercise.

Exercise 6. Show that $|f| \in H^1_{\mathbb{R}}$ implies that $f \in H^1_{\mathbb{R}}$. Hint: Use the preceding exercise and Exercise 3.

Exercise 7. Show that there are $f \in H^1_{\mathbb{R}}$ such that $|f|$ do not belong to $H^1_{\mathbb{R}}$.

Exercise 8. Show that there is a continuous inclusion $BMO \subset L^p(T)$ for all $p < \infty$.

Exercise 9. Show that if $\phi \in C^1(T)$, then the operator $u \mapsto \phi u$ is bounded on BMO and $H_{\mathbb{R}}^1$.

Exercise 10. Suppose that $h \in H^\infty$. Show that the mapping $b \mapsto \bar{h}b$ is bounded from $BMOA$ to BMO. Is it still bounded if $\bar{h}$ is replaced by h?

Exercise 11. Let $BMO_0 = \{f \in BMO; \hat{f}(0) = 0\}$ and define $H_{\mathbb{R},0}^1$ analogously. Show that $H: BMO \to BMO_0$ and $H: H_{\mathbb{R}}^1 \to H_{\mathbb{R},0}^1$ are surjective.

Exercise 12. Show that S in fact maps $L^\infty(T)$ *onto* $BMOA$.

Exercise 13. Derive Theorem 2.1 from Theorem 2.2.

Exercise 14. Let χ be a smooth function on T, and let $d\sigma = e^{-\chi}d\theta$. Show that BMO is the dual of $H_{\mathbb{R}}^1$ with respect to the pairing

$$\langle f, g \rangle = \frac{1}{2\pi} \int_0^{2\pi} f\bar{g}d\sigma.$$

Exercise 15. Let P be the orthogonal projection $P: L^2(T) \to H^2$ with respect to $\langle \, , \, \rangle$. Show that P maps $L^p(T)$ onto H^p, BMO onto $BMOA$, and $H_{\mathbb{R}}^1$ onto H^1. Moreover, show that H^q is the dual of H^p and $BMOA$ is the dual of H^1 with respect to this pairing. Also show that $P: L^\infty(T) \to BMOA$ is surjective. Hint: Note that $SP = P$ and $PS = S$, and so $S^*P = S^*$ if the star denotes adjoint with respect to $\langle \, , \, \rangle$. Hence $P = S^* + (S^* - S)P$. First, assume that f is in $L^p(T)$, $2 \le p < \infty$ or in BMO. Then $Pf \in L^2(T)$.

Exercise 16. Let μ be a Carleson measure in U. Show that

$$u(z) = -\frac{1}{\pi} \int_U \frac{\bar{z}d\mu(\zeta)}{1 - \zeta\bar{z}}, \quad z \in T,$$

is in BMO and that $\bar{\partial}_b u = \mu$.

Notes

The space BMO was introduced by John and Nirenberg in *Comm. Pure Appl. Math.*, 14 (1961). The John–Nirenberg theorem states that if $f \in BMO$ (defined with L^1 instead of L^2 norms; cf. Remark 1.2), then

$$|\{t \in I; |f - f_I| > \lambda\}| \le C|I| \exp\left(\frac{-c\lambda}{\|f\|_*}\right);$$

see [G]. It implies that the two definitions agree; it also immediately implies that $BMO \subset L^p(T)$.

Fefferman's duality theorem is from 1971; see [G].

The space $H_{\mathbb{R}}^1$ can be defined by *atoms*. An atom is either the function 1 or a function $a(t) \in L^\infty(T)$ that has support on some interval I and such that $\int_I a\,dt = 0$ and $|a| \le 1/|I|$. One can prove that $H_{\mathbb{R}}^1$ is the space of all sums

$$f(t) = \sum \lambda_j a_j(t),$$

where a_j are atoms and $\lambda_j \in \ell^1$, and the norm of f is equivalent to the infimum of all $\sum |\lambda_j|$ over all possible representations of f. This definition has considerable advantages; basically, many results are reduced to checking its validity for a single atom; for instance, it is almost immediate that any $b \in BMO$ defines a functional on $H_{\mathbb{R}}^1$, and also that usual singular integral operators are bounded. This definition with atoms was introduced by Coifman (*Studia Math.* 269–274 (1974)), and has several important generalizations to H^p spaces for $p < 1$ in one and several dimensions in $\mathbb{R}^n$ and on spaces of homogeneous type.

In general, the operator $b \mapsto hb$ is not bounded on BMO if h is only a bounded function. There is a precise characterization (due to Stegenga, *Amer. J. Math.* 98 (1976), 573–589) of the multiplicators of BMO: A function h is a multiplicator if and only if it is in $L^\infty(T)$ and

$$\frac{1}{|I|} \log \frac{1}{|I|} \int_I |h - h_I|\,d\theta \le C. \tag{2.2}$$

The condition is satisfied if h has some regularity (C^1 is more than enough); cf. Exercise 9. An analytic function h is a multiplicator on $BMOA$ if and only if $f \in H^\infty$ and (2.2) holds.

There is no *linear* operator that provides a bounded solution to $\overline{\partial}_b u = \mu$ for each Carleson measure μ; however, Exercise 16 shows that the Cauchy integral at least gives a solution in BMO.

Bibliography

[A1] L.V. Ahlfors, *Complex Analysis*, 2d ed., McGraw-Hill Book Company, New York, 1966.

[A2] _____, *Conformal Invariants*, 2d ed., McGraw-Hill Book Company, New York, 1973.

[B] R.P. Boas, *Entire Functions*, Academic Press Inc., New York, 1954.

[D] P. Duren, *Theory of H^p Spaces*, Academic Press Inc., New York, 1970.

[F] G. Folland, *Real Analysis*, John Wiley & Sons, New York Chichester Brisbane Toronto Singapore, 1984.

[G] J. Garnett, *Bounded Analytic Functions*, Academic Press, 1981.

[Hi] E. Hille, *Analytic Function Theory*, vols. I & II, Ginn and Company, Boston, 1959 & 1962.

[Hö] L. Hörmander, *The Analysis of Linear Partial Differential Operators I*, 2nd ed., Springer-Verlag, Berlin Heidelberg New York, 1990.

[N] N.K. Nikolskii, *Treatise on the Shift Operator*, Springer-Verlag, Berlin Heidelberg New York Tokyo, 1986.

[Ra] T. Ransford, *Potential Theory in the Complex Plane*, Cambridge University Press, Cambridge, 1995.

[Ru1] W. Rudin, *Real and Complex Analysis*, 3rd ed., McGraw-Hill Book Company, New York, 1986.

[Ru2] _____, *Functional Analysis*, McGraw-Hill Book Company, New York, 1973.

[S]   E.M. Stein, *Singular Integrals and Differentiability Properties of Functions*, Princeton University Press, Princeton, NJ, 1970.

[S-W] E.M. Stein & G. Weiss, *Introduction to Fourier Analysis on Euclidean Spaces*, Princeton University Press, Princeton, NJ, 1971.

List of Symbols

Index